Beautiful Structures

Beautiful Structures

The Aesthetics of Mathematical Physics and the Art of Creation

CHARLES PICKLES

Foreword by Mark Harris

WIPF & STOCK · Eugene, Oregon

BEAUTIFUL STRUCTURES
The Aesthetics of Mathematical Physics and the Art of Creation

Wipf & Stock
An Imprint of Wipf and Stock Publishers
199 W. 8th Ave., Suite 3
Eugene, OR 97401

www.wipfandstock.com

PAPERBACK ISBN: 979-8-3852-8737-6
HARDCOVER ISBN: 979-8-3852-8738-3
EBOOK ISBN: 979-8-3852-8739-0

VERSION NUMBER 081726

Thank you to Mark Harris—physicist, theologian, and Andreas Idreos Professor of Science and Religion at Oxford University—whose acuity and patience seeded the book this has become, and to the Reverend Dr. Michael Fuller of the University of Edinburgh, who helped draw an inchoate set of ideas into a pleasing structure.

My gratitude also to Dr. David de Pomerai, who patiently endured a reading of my earlier PhD, and to Dr. Jason Frowley, Dr. Neil Harrington, and Jon Convery for their fraternal care and intellectual companionship.

For my sisters, Dr. Jill Planche and Janet Mitchell, and my nieces, Tracy Cocks and Hannah Dunlop.

Contents

List of Illustrations

Foreword

IT IS NOT OFTEN realized in our modern world, but the scientific enterprise has been entangled with concepts of beauty, higher purpose, and divine intentionality since its early years: the so-called Scientific Revolution of the seventeenth century. At that point, the early modern scientists were not shy of calling upon such concepts—in the shape of natural theology—to justify their empirical discoveries in the natural world, and vice versa. Natural theology has a venerable history, going back long before this, perhaps three thousand years or more; its methods, arguments, and claims crop up throughout the world's religious traditions. It is a checkered history though. For, while natural theology was so crucial to the evolution of early modern science, it became deeply unfashionable in the modern Western world after Darwin's discovery of natural selection in the nineteenth century. Natural theology has remained in the doldrums ever since. In recent decades, religious challenges to mainstream scientific method in the shape of the deeply controversial intelligent design movement have only made matters worse. There now seems an impassable gulf between scientific, objective rigor on the one hand, and personal, subjective, and faith-based convictions on the other.

But in the steady march of modern science towards domination of our intellectual landscape, something seems to have been lost, something which the early modern project was able to accommodate with ease by making frequent reference to natural theology. Among many examples that could be chosen from modern scientists and philosophers who lament the loss, witness these musings of the great quantum pioneer, Erwin Schrödinger. The quotation is long (and I have only included the most direct passages), but it is worth reading carefully:

> The scientific picture of the real world around me is very deficient. It gives a lot of factual information, puts all our experience

> in a magnificently consistent order, but it is ghastly silent about all and sundry that is really near to our heart, that really matters to us. It cannot tell us a word about red and blue, bitter and sweet, physical pain and physical delight; it knows nothing of beautiful and ugly, good or bad, God and eternity. Science sometimes pretends to answer questions in these domains, but the answers are very often so silly that we are not inclined to take them seriously. . . . And the reason for this disconcerting situation is just this, that, for the purpose of constructing the picture of the external world, we have used the greatly simplifying device of cutting our own personality out, removing it; hence it is gone, it has evaporated, it is ostensibly not needed. . . . This is the reason why the scientific world-view contains of itself no ethical values, no aesthetical values, not a word about our own ultimate scope or destination, and no God, if you please. Whence came I, wither go I? . . . Science is, very usually, branded as being atheistic. After what we have said, this is not astonishing. If its world-picture does not even contain blue, yellow, bitter, sweet—beauty, delight and sorrow,—if personality is cut out of it by agreement, how should it contain the most sublime idea that presents itself to human mind?[1]

In light of Schrödinger's complaint (and there are many others like it, as I said) over the failure of the scientific method to make sense of those values and purposes that are of ultimate importance to the human person, Charles Pickles's book is a revelation.

I remember first meeting Charles in 2015 when he was inquiring about master's study with me at the University of Edinburgh. Even then, Charles was certain that he wanted to look at beauty in the physical sciences, certain that a case could be made for objectifying the aesthetic properties of nature. I also remember that I was skeptical, but equally I wanted to encourage his patent enthusiasm to learn how to test the case. Over the course of the next seven years Charles dug deeper and deeper into the arguments, first completing his master's in science and religion, and then his PhD. Towards the end of the latter, the breakthrough came, the breakthrough which is argued so effectively in this book. I have not come across an argument like it, and I regard Charles's work as a stunningly original contribution to the science and religion field: likely to be long afloat in the stormy waters of scientific naturalism, while it is surrounded by the rotting hulks of previous attempts to ground beauty, value, and purpose. I will not attempt to describe

1. Schrödinger, *Nature and the Greeks*, 95–97.

Charles's argument; he does that perfectly himself in the opening pages. But I will thank him for his persistence, and for allowing me to show my appreciation in this foreword.

Mark Harris
Andreas Idreos Professor of Science and Religion,
University of Oxford

Introductory Overture

There are absolute masterpieces which move us intensely: Mozart's *Requiem*; Homer's *Odyssey*; the Sistine Chapel; *King Lear*. To fully appreciate their brilliance may require a long apprenticeship, but the reward is sheer beauty—and not only this, but the opening of our eyes to a new perspective on the world. Einstein's jewel, the general theory of relativity, is a masterpiece of this order.

—Carlo Rovelli, physicist[1]

Physicists sometimes speak about their theories in an unusual way. The equations that describe the fundamental structure of the world are praised not only for their empirical success but for their elegance, simplicity, and beauty, as in Rovelli's encomium to Einstein. Paul Dirac, the Nobel physicist of the early and mid-twentieth century, is known for his entirely serious suggestion that it is "more important to have beauty in one's equations than to have them fit experiment,"[2] while Albert Einstein himself repeatedly insisted that the acceptable theories of physics must possess a kind of mathematical inevitability and formal grace. Such remarks are not rare curiosities. Again and again the most powerful descriptions of nature have displayed a striking feature: the deeper the mathematical structure of the theory, the more often it appears to scientists themselves as elegant, harmonious, or beautiful. Yet this way of speaking sits somewhat awkwardly alongside a dominant philosophical picture of the modern world. As a consequence, despite the superficial appearance of these judgments as referring to actual aesthetic properties of the physical world, they are commonly interpreted

1. Rovelli, *Seven Brief Lessons*, 3.
2. Dirac, "Physicist's Picture of Nature," 47.

as nothing other than expressions of the pleasure scientists thoroughly immersed in their disciplines feel when a problem is resolved.

This is not a surprise. Modern thought is accustomed to dividing the world in two. On the one hand lies the physical universe: law-governed, measurable, mathematically describable, and, in itself, supposedly devoid of value. On the other hand lies the evaluative domain: beauty and ugliness, fittingness and distortion, meaning and significance, all treated as additions supplied by human consciousness to an otherwise neutral reality. That division has become so familiar that it is often taken to be little more than common sense. Science deals in facts; value belongs elsewhere. For many, the whole progress of scientific disciplines lies in respecting and defending this perceived dichotomy against the risk of introducing superstition and irrationality back into our understanding of the natural world, or even speculative metaphysics, the latter causing a shudder of horror in some scientific minds.

Yet this familiar picture stands in tension with an obvious feature of our experience of the world: rational creatures do not merely observe nature but purposefully transform it. We reshape the physical environment into artifacts—bridges, buildings, instruments, musical compositions—whose organization is purposive and whose success or failure may be judged according to standards internal to their form. A bridge may be well designed or badly designed; a symphony, balanced or awkward. Even our speech depends on the formation of sounds into designed acoustic shapes that can bear syntactic and semantic properties. This prompts a question rarely raised: Why *should* the physical world be amenable to the rearrangement of its causal properties into forms that are intentionally constructed to serve the purposes of rational creatures like ourselves?

The status of human artifacts provokes a more fundamental question about the character of the world in which such artifacts can arise. For anything that exists, either purposive items, like bridges, and language, or natural objects and events, like atoms and stones arising purely from causal events of the physical world alone, must have structures of some kind. A thing without structure would not be a thing, nor could it have internal relations, no determinacy, and no identity in virtue of which it could be the kind of thing it is. But structure is not a neutral notion. To speak of a structure is already to speak of proportion, relation, balance, hierarchy, tension, integration, or failure thereof. A structure may be coherent or incoherent, elegant or clumsy, balanced or distorted, unified or fragmented, and harmonious or discordant. All of these, and many more besides, involve judgments of value and aesthetics. They are not ornamental predicates added after the fact. They belong to what it is for something to be a structure in the first place.

This observation forms the starting point of the argument developed in this book. If anything that exists must possess a structure, and if structures necessarily admit of evaluative characterization, then the familiar separation between fact and value becomes unstable. Wherever there is structure there is already the possibility of aesthetic judgment because structures admit of better and worse realizations of relational order. The aesthetic dimension of reality, in this sense, is not an anthropomorphic projection onto otherwise indifferent matter. It belongs to the world from the metaphysical start.

It is against this background that the reflections of physicists about the beauty of their theories acquire their full significance. The equations of Maxwell uniting electric and magnetic waves, electromagnetism, the field equations of Einstein, and the symmetry principles of modern particle physics are admired not simply because they work but because their form displays an economy, proportion, and inevitability that invite aesthetic recognition. Dirac's remark about beauty and Einstein's confidence in elegance were therefore not merely personal idiosyncrasies. They express a recurrent feature of the scientific enterprise itself.

Such remarks as noted above are often treated as colorful metaphors serving the purpose of describing a psychological state. Yet the persistence of aesthetic language in the practice of science invites a deeper interpretation. If the structures uncovered by physics repeatedly display features such as symmetry, balance, and simplicity, then aesthetic judgment may be responding to something intrinsic to those structures themselves. If this is so, then exclamations of beauty are not accidental accompaniments to scientific discovery but one of the ways in which the order of reality becomes intelligible, and why base physical structures might even need to be beautiful to work, rather than otherwise. Not only might these remarkable consequences flow from the possibility that the aesthetic properties structures are real, but such a recognition invites, like a work of art, the spontaneous recognition of intentional value and purpose in their origins.

Thus possibility is reinforced by another associated phenomenon that has long puzzled philosophers of science: the application of abstract symbols to the world. The physicist Eugene Wigner once famously described the effectiveness of mathematics in physics as "unreasonable."[3] Why mathematics, developed in abstraction from the empirical world, repeatedly turns out to describe physical reality with remarkable precision seems to be a matter of serendipity. Yet this correspondence is so accurate that without mathematics and accompanying theories, there would be no physics. One only has to think of Newton's relatively simple equation for describing the attractive

3. Wigner, "Unreasonable Effectiveness," 3.

forces operating between bodies that we call "gravity" to understand the power of these abstractions. In addition to these properties of mathematics, physicists and mathematicians almost all find occasion to comment on the aesthetically compelling nature of some equations and their proofs. The most powerful theories, in fact, frequently exhibit precisely those qualities—unity, symmetry, and economy—that we also recognize as marks of beauty.

At this point the philosophical and theological stakes begin to emerge together. If the world is structured in ways that are intrinsically open to aesthetic judgment, and are describable in symbolic "languages" of mathematics, then intelligibility and value cannot be as sharply separated as modern naturalism has supposed. And if the deepest structures of the world exhibit not only order but a kind of splendor proper to their form, then the question of their ultimate ground arises with renewed force.

The theological trajectory of this book therefore does not arise as an extraneous addition to a philosophy of science already complete in itself. Rather, it emerges from the metaphysical implications of a certain scientific picture of the world itself. A universe whose fundamental structures are mathematically intelligible and aesthetically ordered is not easily understood as the outcome of brute fact alone. The suggestion pursued in what follows is that such a world may be more adequately interpreted as the expression of a deeper rational order.

So, to rehearse the discussion so far, the argument proceeds from a simple but far-reaching proposition: anything that comes into physical existence must possess a structure capable of mathematical description.

"Structure" is not an accidental, contingent, feature of things but a necessary condition of their being. Whatever exists must exhibit determinate relations among its parts or properties, and such relations are precisely the kinds of relations that mathematics articulates. If this is correct, then the aesthetic character frequently attributed to mathematical structures in physics is not merely a subjective reaction to successful theories. It reflects a deeper feature of reality itself: the fact that the relational order instantiated in the world admits of evaluation in terms of coherence, proportion, and elegance.

Mathematics articulates relations; structures instantiate those relations; aesthetic judgment recognizes their coherence or failure. Mathematics, structure, and beauty therefore belong together as different aspects of relational order.

Within the conceptual framework that has dominated modern science, the place of artifacts is far from straightforward. The success of the mathematical sciences has depended upon a methodological exclusion of intention, purpose, and final causation from the explanation of natural phenomena. Physical events are described in terms of efficient causes, differential equations, and lawful relations between measurable quantities. Within such a picture the world appears as a vast system of impersonal processes unfolding according to mathematical regularities. Artifacts, however, do not seem to belong comfortably within this scheme. They are objects whose organization is intentionally directed toward ends: a bridge exists in order to span a river, a musical composition in order to realize a form, a scientific instrument in order to disclose aspects of nature. In them, purposiveness is not merely projected by observers but built into the very identity of the object. The existence of such structures therefore introduces into the physical world a dimension that the explanatory strategy of modern science has largely set aside. What appears methodologically excluded at the level of scientific description reappears, stubbornly, in the reality of human making.

Once this point is recognized, the relation between natural structures and human artifacts acquires new significance. The structures described by science do not appear only in nature. They also appear, often in intensified form, in the products of human making. Artifacts, too, like entirely physical objects and events, must obey laws of nature and structure in certain vital respects to have any existence. The well-known Escher paintings play on this fact; they create illusions that impossible structures are possible. Artifacts are structured objects, nevertheless, whose organization is explicitly teleological: their parts are arranged in ways that succeed or fail relative to an intended end. In artifacts the evaluability of structure becomes especially visible because the relation between form and purpose is explicit.

The existence of artifacts therefore raises an important philosophical question. If the natural world were nothing more than a system of impersonal physical processes, it would be difficult to explain how purposive structures could arise within it. Yet human beings, who are themselves part of the natural world, continually reorganize physical materials into artifacts whose identity depends on intention and purpose. The world therefore appears capable of accommodating teleology as well as causality.

These reflections place pressure on the naturalistic picture of reality that has dominated much modern thought. According to ontological naturalism, the ultimate constituents of the universe are exclusively physical processes and entities. Teleology, normativity, and value arise only later as by-products of biological evolution or human cognition. On this view aesthetic judgment is a projection onto a fundamentally indifferent world.

Such a picture has been sustained in part by an important methodological principle that governs the modern sciences. Scientific explanation proceeds by seeking lawful relations among observable phenomena and by avoiding appeal to purposes, intentions, or transcendent causes. This methodological restraint has proved enormously fruitful. Yet what began as a practical rule for conducting empirical inquiry has often been transformed into a far stronger metaphysical claim about the nature of reality itself. The success of the scientific method has encouraged the belief that the world must ultimately be exhaustively describable in the terms employed by the natural sciences.

The transition from methodological naturalism to metaphysical naturalism is rarely argued for explicitly. Instead it tends to occur gradually, as explanatory practices become reinterpreted as ontological commitments. Because the sciences explain phenomena without immediate reference to teleology or divine agency, it is assumed that such realities do not exist. Yet the logic of this inference is far from straightforward. A method designed to isolate certain kinds of causal relations cannot by itself determine the full character of reality.

Indeed the very possibility of scientific inquiry presupposes conditions that cannot easily be captured within a purely naturalistic framework. Scientific reasoning depends upon norms of rationality, standards of evidence, and commitments to truth that transcend the descriptive vocabulary of physics or biology. Moreover, the intelligibility of the world itself—the fact that nature exhibits stable structures capable of mathematical articulation—is something science presupposes rather than explains.

From this perspective the beauty of the mathematical structures uncovered by physics takes on a new significance. If aesthetic judgment tracks the coherence and intelligibility of relational structures, then the beauty recognized by physicists may reveal something about the character of the world itself. The aesthetic dimension of scientific practice may therefore point beyond methodological convenience toward a deeper metaphysical reality.

This suggestion opens a path toward a renewed dialogue between science and theology. The modern conversation between these disciplines has often been dominated by questions about cosmological origins or the fine-tuning of physical constants. Such discussions, while important, tend to frame the relation between science and religion in terms of explanatory gaps or probabilistic design arguments. The approach pursued here proceeds along a different route.

Rather than focusing primarily on cosmological contingencies, it asks a more basic question about the structure of the world itself. Why should the universe be intelligible through mathematics at all? And why should the structures revealed by physics so often display forms that scientists themselves describe as beautiful? If the aesthetic dimension of structure belongs to the very conditions of intelligibility, then the beauty of the physical world may be understood not as an accidental feature but as an expression of its deepest order.

Within the Christian intellectual tradition, beauty has long been understood as a transcendental of being. From Augustine through Aquinas to the theological aesthetics of Hans Urs von Balthasar, beauty has been regarded as the radiance of form—the manifestation of order, proportion, and intelligibility within the fabric of reality. On this view the experience of beauty is not merely subjective delight but a recognition of the coherence of being itself.

If this tradition is taken seriously, the beauty encountered in the mathematical structures of physics may be interpreted as a disclosure of the rational order underlying the cosmos. Mathematics would then articulate the grammar of relations instantiated in the world, while aesthetic judgment would recognize the coherence of those relations. The intelligibility and beauty of the universe would thus appear as complementary aspects of a deeper metaphysical unity.

Such a perspective also illuminates the place of human creativity within the natural order. Artifacts are not arbitrary impositions upon a neutral world but participations in the same structural logic that governs nature itself. When architects design buildings, composers write symphonies, or engineers construct bridges, they organize materials according to relations that can succeed or fail in ways that are objectively intelligible. The evaluative standards applied to artifacts arise from the internal logic of the structures themselves.

Human creativity therefore reflects a remarkable feature of the world: the same relational principles that govern natural structures can be formed into new structures through purposive agency. In this sense artifacts disclose something about the compatibility between the intelligibility of nature and the rational capacities of human beings. The world appears hospitable to minds capable of recognizing and reproducing its patterns.

These reflections point toward a metaphysical vision in which mathematics, structure, beauty, and teleology are intimately connected. Mathematics articulates the relations through which structures become intelligible. Beauty marks the successful realization of those relations. Teleology represents the purposive organization of structure toward intelligible ends. Together these dimensions suggest a world whose order is not merely mechanical but expressive.

The thesis advanced in this book is that the aesthetic character of the mathematical structures revealed by physics provides evidence of an underlying rational order in reality. The world investigated by science is not only lawful and intelligible but displays a coherence that invites aesthetic recognition. This convergence of intelligibility and beauty is most plausibly understood if the structures of the cosmos are grounded in a deeper rational source.

That the world necessarily comes in structural and aesthetic forms can be quite sufficient to provoke theistic inclinations: the uncovering of expressed beauty at the empirical heart of the physical manifestation of creation, however, conveys transcendent emotions and thoughts which objects of beauty often provoke; they convey something of themselves but something of their creator beyond them. As philosopher Anthony O'Hear puts it,

> The religious impulse [is] a response to a natural object or work of art as beautiful: in which we see a thing not just as a . . . "fragment of nature," but something which mediates between our own longing for perfection and some other world in which that perfection is realised.[4]

If this is so, the beauty of the mathematical structures discovered by physics may be understood as more than an aesthetic curiosity. It may be one of the ways in which the world discloses its ultimate source. The suggestion pursued here is that the structures of the cosmos bear the marks of intention in much the same way that human artifacts do. The world, in this sense, may be understood as the work of a supreme artificer—an *artifex pulcherrimus*—whose creation is not only intelligible but beautiful. The significance of the experience of beauty in the swirling complexities of mathematical physics, however, is not exhausted by the psychology of the moment. In the intricate but purposeful beauty of the mathematical structures at the base of the physical world, we perceive what we recognize in our own: a personal intent that it should be so.

Such a claim does not seek to replace scientific explanation with theological speculation. Rather, it proposes that the success of scientific inquiry itself presupposes a metaphysical context in which intelligibility, rationality, and value already have a place. Taken together, these reflections aim to articulate a vision of reality in which the mathematical intelligibility of the universe, the beauty of its structures, and the purposive creativity of human artifacts are not disconnected phenomena but expressions of a common underlying order. The wager of this book is that by attending carefully to these convergences we may glimpse something of the deeper rationality that grounds the world itself. The aim is therefore not to challenge the

4. O'Hear, "Science and Religion," 512.

achievements of modern science but to explore the philosophical and theistic implications of those achievements.

Yet the inquiry pursued here extends further. The second half of the book turns to the more difficult question of whether such a world may rightly be understood as personal in its ultimate source. This requires more than a simple analogy between human artifacts and nature. It requires reflection on the status of teleological perception itself: whether our tendency to discern purpose in natural forms is a genuine mode of recognition, or merely an evolved habit projected onto indifferent processes.

For that reason, the later chapters examine the strengths and limits of natural theology, the role of deductive arguments for God, and the question whether either naturalism or theism can claim a privileged position of epistemological neutrality from which the whole character of reality may be judged. The issue is not simply whether one can construct a proof from the world to God but whether the world is more truthfully seen when its structures, beauty, purposiveness, and emergence of personal life are apprehended as creation rather than as the accidental outcome of impersonal process alone.

This also includes our own place within the world. We are not merely observers standing outside nature and drawing analogies from artifacts to God. We are ourselves embodied, rational, purposive creatures within the order we seek to understand. Our capacity to recognize form, create artifacts, pursue ends, and apprehend meaning may itself be understood as a finite participation in the rational and personal source of that order. In this sense the question of divine personhood is inseparable from the question of how personhood, intention, and rationality are disclosed within creation itself.

OVERVIEW OF THE ARGUMENT

Part 1

In defense of the position that beauty in intelligible structures reflects divine intention, I begin by providing an alternative to what is generally called an "ontological naturalist" picture of the universe and all that it finally consists of. There are many variants of what the ultimate occupants of a naturalist universe are said to be. Some incorporate abstract elements like mathematics into their fundamental ontologies; others firmly deny anything other than what a physical world, in all its natural forms, provides. Some allow that higher-level phenomena may emerge in ways not reducible in practice to microphysics, while still maintaining that nothing finally exceeds the natural order itself. Others are more austere, holding that whatever cannot in principle be rendered within a completed physical account has no serious claim upon reality.

The essence of the picture is that all that ultimately can be said to exist consists of the immanence of the "material" or the "physical" in one way or another. No signs of divine life, intelligence, or inkling of the personal is to be discerned in that ontological space, and neither are any other signs of teleological life except, possibly, as outgrowths of the physical in biology. Whatever appears purposive, meaningful, normative, or expressive must, on this view, either be reconstructed in nonteleological terms or treated as derivative appearance generated by more basic impersonal processes.

I present the naturalist case as it is normally made with the presence of physics in the background or the foreground because of its scientific prestige. It is the discipline which is widely accepted as investigating the most fundamental physical structures on which all else that is "natural" must ultimately depend, however the relationships between other various disciplines of the natural, like chemistry or biology, and the "physical" in physics are to be understood. Whether the higher sciences are reduced, realized, emergent, or otherwise related to physics, the prestige of the physical as explanatorily basic commonly remains in place.

Naturalist pictures of the cosmos all begin where the practices of contemporary sciences begin, by extruding any reference to the personal and the divine as a point of *methodological* principle:[5] explanations of atomic, molecular, inanimate, and organic processes are forbidden reference to teleology, both human and divine, in any theory that purports to explain natural phenomena. This methodological discipline has obvious and genuine power. It guards against premature appeal to hidden purposes, secures testability, and directs inquiry toward publicly examinable causal relations.

It is a principle supposedly free of any metaphysical commitments. Nevertheless, it is frequently inflated into a metaphysical hypothesis about the nature of all reality: everything that exists or could exist is either reducible to, or dependent on, the physical. The essential argument here is that the extraordinary progress of science, achieved by restricting explanation to "external" physical processes and causation, supports the metaphysical extrapolation. What began as a rule of method is thus converted into a doctrine of being.

5. Whether in fact the idea of a such a principle does detailed useful work in understanding the actual practices of science is an interesting question. I use it here, however, because it features as an important philosophical tool in supporting the metaphysical view I am contesting.

The import of inflating the methodological principle into an ontological criterion is to deny any role to purposes, intentions, and appeals to any broader understanding of physical phenomena in terms appropriate to, for example, the aesthetic creations of human persons. While one might say, "The deep mathematical physics of the world is beautiful," this has no serious ontological implications for the ultimate nature of all things other than as an observation of how it appears to us. Beauty, on such a reading, is relocated entirely into the spectator and stripped of revelatory significance.

One result of the primacy assigned to the idea of the physical in this picture results in sharp dichotomy being drawn between the "personal" and the "impersonal."

The "impersonal" world, that of the physical and its causal processes, is promoted to a position of ontological hegemony: physical causes and causal regularity are what ultimately exist or have "real" existence. They are taken as self-sufficient, requiring no appeal beyond themselves in order to be what they are.

Teleology, persons, normative standards, and consciousness are assigned a secondary status: whether the causal result of physical processes, preferably reducible to them, or at least supervenient on them, all these phenomena are conceived to be both temporally and logically dependent on the prior existence of the "physical." Their reality is tolerated, where it is tolerated at all, only under license from a more fundamental impersonal order.

One way of understanding this claim is to see it as entailing that before the evolution of the organic life of sentient and sapient creatures, there were no values—epistemological, aesthetic, no normative standards against which anything might be evaluated—and no teleology: the universe was simply a churning activity of the "physical," whatever a final theory of physics will determine that state to be. There were facts, perhaps, but no truths; arrangements, perhaps, but no meanings; processes, perhaps, but no reasons.

A natural consequence of finally seeing nothing but empty physicality in the universe has led to it being described as "disenchanted"; no meaning, purpose, or teleology is to be found among its basic constituents, and hence, it cannot manifest any expressive properties like beauty suggestive of divine teleology and personhood. At best, enchantment becomes a human projection cast upon an otherwise indifferent base reality.

I maintain that contrary to the naturalist picture, there is an inclusive and overarching metaphysical framework of the world which provides a place for personhood and its artifacts, "physical" processes and events, scientific methodology, normativity, and teleology from the ground up as equal ontological participants. My contention is not that causal explanation is false but that it is partial: one legitimate register within a richer account of what reality contains and what must already be the case for causal explanation itself to arise.

Contra the physicalist naturalist, I argue that the idea of an external world scientifically explicable via empirically identifiable causes and the subsequent generation of laws of nature necessitates a prior inclusive framework. As Bruce Gordon maintains, "What science is depends largely on how human beings understand the nature of nature itself."[6]

"Science" thus cannot provide an explanation of events and relations understood as causal and regular since these are conditions of the emergence of what a "scientific" investigation of the natural means. The critical idea of evidence in science, for example, necessitates a normative epistemological relation between physical structures and cognitively sapient creatures who can exploit that possibility to gather knowledge. In other words, we must have a grip on what is to count as "evidence," "truth," and their relations before science is possible. Rules of inference, standards of relevance, discriminations between signal and noise, and the authority of reasoned correction are not discoveries of laboratory method but preconditions of it.

Yet the claim I wish to press is stronger still. It is not merely that science presupposes norms of inquiry once a world is already given. Rather, something like a world must already stand forth as intelligible before scientific practice can begin at all. To speak of objects, events, regularities, causes, measurement, or confirmation is already to inhabit a space in which reality is available under concepts. As philosopher John McDowell has stressed in another context, minded creatures do not first confront a bare manifold and then impose sense upon it from without; the world comes to us already within the reach of understanding.

If this is right, then normativity is not an optional addition supplied by later human reflection, but it belongs to the very possibility of there being items that can count as anything for us. An occurrence must be determinable as this happening rather than that; a thing must exhibit identity, boundary, persistence, relation, and difference if it is to be encountered as an object at all. The contrast between what is relevant and irrelevant, fitting and unfitting, same and different, true and false, is woven into the disclosure of a world before it becomes the subject matter of experiment. We do not first receive brute data and then later decide what shall count; the possibility of counting already presupposes order.

The overall metaphysical alternative I develop thus includes in its scope the idea that teleological and normative properties play a role elucidating

6. Copan and Reese, *Three Views*, 192.

what a world fit for the methods of science must look like. It also includes the stronger thought that the intelligibility through which a world first appears cannot be detached from standards internal to what counts as order, coherence, and determinacy. A universe wholly devoid of such dimensions would not merely be unknown; it would be unworlded.

My principal focus will be to assert the metaphysical proposition that anything that comes into physical existence must have a mathematically describable structure.

I should stress the metaphysical range of this proposition. It is a prescription about the logico-mathematical principles that are necessary in order for anything to be counted as existing at all, in whatever form. I explain this fully later on. It does not prohibit the individuality or diversity of life. In fact, it is a prerequisite of being picked out at all as an "individual thing" in the natural world that it shares general features of existence that make it so determinable. To exist concretely is already to be available to distinctions of one and many, continuity and discontinuity, relation and measure.

I will argue in tandem with this that anything that has a structure necessarily has an aesthetic characterization. By this I do not mean that everything is beautiful in some consummate sense but that wherever there is structure there are already questions of proportion, coherence, fit, symmetry, tension, completion, or defect. These are not merely decorative predicates added after perception; they belong to the ways in which anything structured can succeed or fail in being what it is. If so, aesthetic normativity enters not only when we admire works of art or elegant theories, but wherever anything counts as an object or event at all.

Because aesthetic judgments are normative—i.e., refer to some standard external to the thing being judged—this proposal, if true, has the profound consequence of implying that the origin of any physical world must have, in direct opposition to naturalist assumptions, at least one normative structure, an aesthetic one. If no structured reality can be wholly indifferent to standards of fit and defect, then normativity cannot be a late accidental bloom upon a valueless stem.

What the aesthetic character of any extant structure is, however, is an a posteriori matter. The uncovering of physics (for which I provide evidence), that the mathematical structures at the base of the natural world are beautiful, provokes the question as to why it has one aesthetic form rather than another—a normative enquiry.

Part 2

Reflection on the lives of human persons opens a way to answering this question. We, too, are a part of the "nature" founded on the same physio-mathematical base. We are manifested as corporeal creatures by it and engage with it via rational and teleological capacities that allow us to purposefully create artifacts that are not the outcome of regular causal processes. Contra physicalist ontological naturalism, "nature," I argue, must therefore provide a place for teleology that allows for the purposeful creation of novel items that are not the products of "laws of nature." If beings continuous with the natural order can genuinely originate purposive constructions, then teleology cannot simply be an illusion cast by blind mechanism.

The argument I offer is not that there are entities that defy naturalistic explanation in some respect but rather that anything that is to come into existence will have to obey the mathematical and structural characteristics that are necessary to its being the thing it is. Our artifacts, after all, in order to come to fruition must obey the same normative mathematics and aesthetic principles as any "physical" event or object. Nevertheless, our capacities to construct items whose history, purpose, and functions can only be fully comprehended in teleological terms is prima facie evidence that there is more to the natural reality in which reflective and teleological creatures like ourselves move than can be accommodated by a reductive naturalism. There is nothing "unnatural" about the physical composition of a violin, but its identity, use, meaning, and history are teleologically dependent. To know what it is requires more than inventorying its materials or charting its efficient causes.

In his discussion of the relation of the human creator to his work, the art theorist Aaron Ridley maintained that "the artist must be seen as present in his work, much as a person must be seen as present in his behaviour, rather than as separate from it, behind it, or, above all, logically independent of it."[7]

Ridley's point covers not only artistic artifacts but all human creations, of which artistic items are but a subset; every physical item whose contours are the outcome of purposeful intention presents their humanity in their very construction. Intention is not externally attached to such things as a detachable label; it is sedimented into form, arrangement, relation, and use. The maker is encountered in and through the made object.

It is important to be aware of the scope of such things, which range from the shaping of sounds into conveyors of meaningful speech, to flint

7. Ridley, "Expression in Art," 218.

axes, the organization of electromagnetic waves for purposes of control and communication, and the tools of experiment; hadron colliders, telescopes, even the shaping of movements in a ballet are all intentional acts. Civilizations become visible in what they build, preserve, refine, and hand on. A culture's intelligence is objectified in its artifacts.

It is sometimes supposed that because some natural objects look as if they are the outcome of intent, but are, in fact, entirely shaped by ordinary causal events, that this throws the distinction between teleology and the efficient causation of natural events into doubt. Not so. Resemblance of outcome does not erase difference of source. A crystal may display symmetry, and a river delta branching order, without thereby becoming equivalent to a violin sonata or a bridge.

For the theist, the very presence of the ordered physical world is a divine creation. Human intentionality is but a pallid sharing of that divine teleology. So a distinction and recognition of the objects we create within the world is a recognition of the participatory but severely limited power of construction we are granted as creatures within a greater cosmic artifact. Our making is derivative, local, and finite, yet real.

Other creatures, too, create artifacts—the bee a hexagonal comb, the beaver a dam—but as we are restricted, so, too, is the intentionality of these creatures comparative to our own. Nevertheless, we see the commonality between our creations and their own without that recognition threatening the different modes and extent of their artifactual powers relative to our own. Degrees of purposive agency may differ without collapsing into identity or disappearing into mechanism.

Beauty in our artifacts is an attempt to give expression to something existentially pleasing in the creative ordering of the physical in certain ways. Part of this experience is the pleasure we take in manipulating the material world to powerfully elicit the aesthetic intensity that resides within it, and to which we so profoundly respond. We delight not merely in achieved form: in balance, resonance, elegance, aptness, and grace.

This, I will suggest, is analogous to us perceiving in the aesthetics of the structures uncovered by mathematical physics the intentional hand of a creator. That sense is enhanced by the contingent presence of beauty in physics. Here, for example, are the words of Werner Heisenberg shortly after developing his matrix in 1925 that gave the numbers describing the "leaps" of an electron from one energy level to another inside the atom. The development of this work was part of the beginning of quantum physics, a development that profoundly altered our picture of reality. The older image of matter as composed of minute objects possessing fixed properties and moving in wholly continuous and determinate ways could no longer be straightforwardly sustained. At the most fundamental level, nature increasingly disclosed itself through probabilities, discontinuities, and relations that resisted ordinary visualization.

> I had the feeling that I had gone beyond the surface of things and was beginning to see a strangely beautiful interior, and felt dizzy at the thought that now I had to investigate this wealth of mathematical structures that Nature had so generously spread out before me.[8]

It is striking that this decisive advance was achieved not by clearer sensory pictures of what matter "looks like" but by mathematical articulation of the lawful structures governing its behavior. What the underlying entities finally are became more elusive just as the formal relations describing them became more exact. In that sense, mathematics did not merely accompany discovery here but became the primary medium through which the deep constitution of the physical world was grasped.

The world, after all, could have been mathematically impenetrable, let alone obeying the dictates of unity, simplicity, and symmetry at the center of physics. The thought that the physical world does, in fact, have at its center not only order but principles coincident with great beauty, reinforces the idea of normative intent behind and in it. It is not only that there is lawfulness but that lawfulness repeatedly appears under canons long associated with intellectual and aesthetic excellence.

Like all beauty, it can produce a longing, often inarticulate, that seems to be neither fully contained in nor fully satisfied by the object which provokes it, a desire for something that exceeds the world in which the object and the observer participate. Beauty commonly intimates surplus: more significance than finite form seems able fully to contain.

When the thought of the basis of the material world as an intentionally constructed artifact is combined with the thought that any world expressing teleology is hard to make sense of by an entirely nonteleological reality, the idea and experience of an encounter with the world as an outcome of divine creation becomes even more plausible. Like our encounter with other persons, however, there is no final deductive proof. I suggest, however, that this is no obstacle, either rationally or existentially, to an experience of the world as divine artifact. Many of the most important realities we acknowledge are known by modes other than demonstration alone.

We ourselves may be understood within this picture as created participants in that same artifactual order. As embodied rational creatures, capable

8. Heisenberg, as quoted in Rovelli, *Helgoland*, 13.

of intention, judgment, and purposive making, we do not stand outside creation as detached observers but realize our own nature within it. Our consciousness and creative agency are themselves among the ways in which the deeper intelligibility of the world becomes manifest.

Few people, I imagine, have ever seriously entertained the idea that the ceiling of the Sistine Chapel is entirely the outcome of nondirected physical events on the grounds that there is no evidence in the physical picture itself for the person of Michelangelo (or any person, to be more precise; or, to hone that precision even further, that there could be, the painting itself being nothing other than a conglomeration of complex physical arrangements). We ordinarily take purposive form to be revelatory of mind because, in many contexts, it plainly is.

No doubt, in the far reaches of the philosophical debate, many naturalists have argued that there are, really, no such phenomena as teleological objects and personal creators, and that ultimately, in the distant future of "science" and philosophy, all such claims will give way to reductive analyses of a strictly efficient causal kind in material terms. My suggestion is that this confidence often depends less upon discovered fact than upon prior metaphysical convention.

In pursuit of my goal, I examine whether or not there is any position of epistemological neutrality from which the issue could be finally decided. Neither the naturalist nor the theist, I suggest, can claim from within the world a view that confers this cosmic authority. All judgment proceeds from within some understanding of what reality is like and what kinds of explanation are admissible.

I nevertheless deny that this is a particular problem for theists. There is no deductive certainty that those I love are persons. The idea, however, that their corporeality is an obstruction to my certainty that they are so, and that what is required is a deductive proof relating their "mind" to their "body" rather than the latter being a means of the expression of fleshly personhood, is the product of the idea that there is a "mind–body" problem, a philosophical view partly bequeathed to moderns by René Descartes (1596–1650). Yet in ordinary life, embodied and rational presence is precisely how personhood is known.

Analogously, the idea that the physicality of nature requires a deductive proof relating the "person" of God to its manifestation to count as "knowledge" might similarly be thought to be the expression of a particular philosophical view of reality. Conceiving "natural" reality aright may make the presence of God transparent in it. As I show below, at least two theologian-philosophers have argued this to be the case.

A MODEL OF INTEGRATION

This is clearly a particular model[9] of how one is to understand relations between theism, aesthetics, and the "natural" in mathematical physics. It might loosely be called an integration view.

I take the meaning of this term as I use it from the philosopher and historian of the philosophy of science Bruce Gordon. His position is articulated in a three-way dialogue between himself, Alister McGrath, Idreos Professor of Science and Religion at Oxford University, and Michael Ruse, retired professor of philosophy at Florida State University in philosophy of science and evolutionary biology.

In the 2021 book in which the debate occurs,[10] Ruse argues for the view that science and religion represent two autonomous and independent fields, as a consequence of which there can be no conflict between them. McGrath insists that "dialogue" between theism and science can be mutually beneficial. Gordon, while closer to McGrath, is suspicious of the view. "Dialogue," he writes,

> is nice, but it only takes you so far. Moving beyond . . . notions of mutually enriching conversations to a robust metaphysical integration of science and philosophical theology is necessary.[11]

The book is an attempt at such an integration turning on the metaphysical pole of a necessarily aesthetically structured universe, the contingent finding of beauty in the formation of its fundamental building blocks, and the response to these structures as reflecting the hand of a creator who intended it to be so: a proper recognition of the world and all its constituents and structures, abstract, physical, personal, and teleological, as the creation of an *artifex pulcherrimus*, a divine artificer.[12]

INTEGRATION AND THEOLOGICAL AESTHETICS

Although the argument developed in this work proceeds largely through analytic philosophy, philosophy of science, and reflection on mathematical structure, beauty, and artifacts, some readers may recognize that its

9. Barbour, *Myths, Models, and Paradigms*, is a locus classicus for different models on the relation between "science and religion." See also Stenmark, *How to Relate Science and Religion*, and Southgate, *God, Humanity, and the Cosmos*.

10. Copan and Reese, *Three Views*.

11. Copan and Reese, *Three Views*, 16.

12. Literally translated "the most beautiful artificer."

conclusions bear a resemblance to themes long present within the tradition of theological aesthetics. I do not rely upon that tradition as a foundation for the argument developed here. Nevertheless, a brief indication of this broader intellectual context may help situate the claims being advanced. By theological aesthetics I mean those theological approaches in which beauty is understood not as a merely subjective response but as disclosive of being, creation, and divine reality itself.

Nevertheless, the integration I have described between the world and all its constituents and structures—abstract, physical, personal, and teleological—as the creation of an *artifex pulcherrimus*, a divine artificer, clearly resonates with this broader theological tradition. The point is not that the argument of this book depends upon that tradition for its validity. Rather, it is that a conclusion reached through reflection on structure, mathematics, physics, aesthetics, and artifacts finds a natural home within a Christian metaphysical vision in which beauty, truth, goodness, and being finally belong together.

Hans Urs von Balthasar (1905–1988), the Catholic philosopher and theologian, provides one of the most important modern theological articulations of this vision.[13] For Balthasar, beauty is not merely one topic among others. It is a transcendental feature of being, disclosed through the forms of created reality. The beautiful appears in concrete things: persons, works of art, natural forms, acts of goodness, and the ordered structures of the world. Yet in appearing there it also points beyond itself. Created beauty is finite, particular, and sensuous, but it bears within itself a splendor that intimates the depth from which it comes. As Balthasar writes, "No metaphysics of Being as such, and its transcendental qualities, can be separated from concrete experience, which is always of the senses."[14]

Beauty is therefore both immanent and transcendent: encountered in the world, yet never exhausted by the world.

This is closely related to Balthasar's distinction between form and splendor.[15] As form, beauty can be grasped in proportion, harmony, relation, number, order, and intelligible structure. It is therefore not surprising that mathematics and physics should become places where beauty is encountered with peculiar force. Yet form is not beautiful merely because

13. Von Balthasar, *Glory of the Lord.*

14. Von Balthasar, "Transcendentality and Gestalt," 34b.

15. The "beauty" revealed in the concrete particulars of experience for Balthasar are manifested, as he writes, in the "form" or essence of the thing and its "splendor." These two elements have, he maintains, "traditionally controlled every aesthetic and which, with Thomas Aquinas, we could term species and lumen . . . form and splendour." Von Balthasar, *Glory of the Lord*, 115.

it is ordered. It becomes beautiful when its order appears as expressive of a depth or fullness not reducible to its material arrangement alone. This is why Balthasar can speak of truth as symphonic. Reality is not rightly grasped when its elements are treated as fragments. ("Die Wahrheit ist symphonisch," "truth is symphonic," is one of Balthasar's favorite expressions.[16]) They must be seen together, as parts of a whole whose meaning exceeds any isolated component.

This thought is directly relevant to the present work. The beauty of mathematical physics, if real, cannot be adequately understood as a detached curiosity within scientific practice. Nor can the beauty of human artifacts be understood merely as a psychological response to complex arrangements of matter. In both cases form and intelligibility appear together. The world is disclosed not as a collection of unrelated physical facts but as a structured order capable of being apprehended, judged, and, in certain respects, loved.

David Bentley Hart's theological aesthetics develops a related vision. In *The Beauty of the Infinite: The Aesthetics of Christian Truth*,[17] Hart opposes accounts of reality in which difference, finitude, and plurality are finally understood in terms of rupture, violence, or domination. Against this, he presents creation as the finite expression of divine beauty and peace. The variety of created forms is not a fall away from unity but a participation in the infinite generosity of God. Beauty is therefore not merely subjective delight but the form in which the goodness of being becomes visible.

Hart's emphasis is important because it resists the modern tendency to separate God from the visible and finite world, as though the divine could only be thought by abstracting from particularity. As Hart puts it, the beauty of the "unrepresentable, the infinite Other, is visibly continuous with the beauty of this world, and further, its beauty has taken a concrete and particular form. This is the truly good fabric of reality."[18]

For Hart, as for Balthasar, the beauty of the world is not a barrier to the divine but one of the ways in which divine reality is disclosed. The finite does not conceal God simply by being finite. It may mediate, refract, and participate in the infinite source from which it comes.

The present book does not attempt to reproduce the theological architectures of either Balthasar or Hart. Its method is different. It begins not with Christology, Trinitarian doctrine, or the drama of revelation, but with the more restricted question of structure: What must be true of any world capable of mathematical articulation, aesthetic judgment, purposive agency,

16. Nichols, *Key to Balthasar*, 9.

17. Hart, *Beauty of the Infinite.*

18. Kantzer Komline, review of *Beauty of the Infinite*, 39.

and artifactual form? Yet the direction of travel is not unrelated. If the physical world is mathematically intelligible; if its structures are susceptible to aesthetic apprehension; if human beings within that world are capable of recognizing, judging, and intentionally forming beauty; and if artifacts disclose the presence of mind in structured form, then the world begins to look less like a closed system of brute impersonal process and more like a created order.

This is the sense in which the argument offered here may be understood as a preliminary contribution to theological aesthetics. It seeks to show, from within reflection on science, mathematics, structure, and human making, why beauty should not be treated as an accidental surface of reality. Beauty belongs to the way intelligible form is disclosed. If that is so, then the theological claim that creation is the work of a beautiful artificer is not an arbitrary imposition upon the world but a way of seeing the world that may be continuous with what its structures already show.

Whether the theological vision emerges as a consequence of the argument, or whether it has structured the intelligibility of the inquiry from the beginning, is itself part of what this work is concerned to examine.

SUMMARY OF CHAPTERS

Chapter 1

A Sketch for Re-Enchanting Nature: Science, Normativity, and Metaphysics

The aim of this chapter has been broadly discussed above. I outline in it an alternative metaphysical conception of the world that unifies the existence of a causally governed natural world with persons, values, and expressive properties. I maintain that the very idea of a "natural" world or "nature" accessible to scientific practice in its various manifestations depends on this overarching picture which shows their various relationships to be part of a basic ontological union between them.

Chapter 2

Structures in Mathematical Physics: A World of Phenomenal and "Measurable" Beauty

In this chapter, I move on to give an account of how the idea of "structure" is a metaphysical one, by which I intend the claim that anything that is to come into existence must necessarily have a "structure" in order to exist at all. Continuing and deepening the case made in the previous chapter, I show how the basic structures of the physical world investigated by particle physics can be properly understood as uniting subjectivity, normative properties, and mathematical measurement without threat to empirical science.

I do this by showing how the relations between structure, mathematical physics, and aesthetics allow for that possibility. While mathematics describes the objective quantitative dimensions of structures, the expressive properties that they manifest aesthetically open them up to subjective judgment. Despite the subjective requirement of the truth, aptness of the judgment is not relative to the person who makes it. I claim that there are different but epistemologically legitimate modes of access to basic "physical" structures, both "personal" and "impersonal."

In other words, while aesthetic judgments of these arcane structures of basic matter can be referred to their mathematical construction, they are also subject to personal judgment. This does not disqualify them from being real features of the physical world. In fact, I note, the idea of structure is intimately linked to aesthetic features of simplicity, unity, symmetry, and so on, and judgments about such things are critical to theoretical evaluations of the rightness of a structural "order."

Chapter 3

Human Artifacts and Teleological Structure

The structure of the physical world is not the only dimension of existence in which structures, objects, and processes appear. We construct them from what nature makes available to us. I point out that the fact that nature "allows" itself to be manipulated to bring about artifacts to fulfill the teleological aims of sentient creatures is not itself a physical law of nature. It is rather a property of a world that can accommodate rationality and teleology.

I suggest that being structures like any other, our artifacts share the same metaphysical principles with all other structures: they also express aesthetic and normative properties. I show that the teleology they manifest is shared with that of "natural" objects whose causal origins do not involve human intention: How else is one to describe the beaver's dam or the spider's web?

However, while biological processes are necessary to the *ability* of conscious rational creatures to engage with the concepts of teleology and act in accordance with them, they are not sufficient to *explain* the ontological provenance of teleological and intentional standards.

The conceptual distinction in explanation between regular physical causation and teleology does not support an ontological dichotomy within the metaphysical framework I have advanced any the more than the distinction between what is cognitively required to play chess and the rules of chess indicate an untraversable dichotomy between game and player.

Chapter 4

Universal Structures of Beauty and Their Transcendental Teleology

Having established the metaphysical foundations of normative properties in the personal and impersonal worlds, particularly aesthetics, in this chapter I discuss the universal character of "beauty," how it contributes to arguments for God's existence, and the relation between that beauty and divine authorship. I begin by noting that since everything has an aesthetic character, it is to be expected that among the very different events and objects that constitute the world, many will be beautiful.

"Beauty" as I have shown is associated with theoretical judgments, artistic evaluations, and intuitions of the "rightness" of any structural order which manifests it. Experiences of beauty in the physical sciences, nature, and human artifacts often produces existential questioning and yearnings among theists and atheists alike, provoked by something in its nature which has a psychologically transcendent effect.

Beauty would thus seem to constitute an important area of interest for natural theology. I offer some evidence to suggest this is surprisingly not the case. One reason may lie in the fact that the appeal to "beauty" as producing transcendental desires is, or can be taken to be, entirely phenomenal as opposed to rational, or "subjective" rather than "objective," assuming that these modes of apprehension necessarily have ontological consequences.

Contrary to this collection of thoughts I suggest that the association of structure, artifact, nature, and aesthetics draws one intuitively to personal intent once the components of the broader framework I have developed are brought into conceptual play.

That an artifact and an object of nature share their ontological heritage weakens the idea that objects of nature cannot display as part of their identity the personal hand of divine creation. It does so because it intimates that a way can be found to integrate teleology, the normative, and the personal into the conditions of the existence of "physical" objects, thereby bringing them within the scope of teleological intent. This having been accomplished, it is a short step to implicate divine intentionality.

The language of divine artistry interpolates itself because it alone can encompass the presence of the personal and the presence of the purely physical and causal in the same metaphysical and ontological landscape that is our world. Given that all the objects of human life instantiate the teleological, when we find beautiful structures fundamental to the existence of all physical being lurking in the aesthetic thickets of the natural, it is a spontaneous inclination to feel the presence of a personal creation.

Chapter 5

A Metaphysically Designed World: Teleology and the Personal in Arguments for God's Existence

In this chapter I aim to defend and apply the notion that teleology and divine personhood lie at the uniting heart of an ultimate conception of the world. Because this idea invokes the concept of intentional design, it can, wrongly construed, appear as an explanatory hypothesis in competition with the assumptions of scientific explanation and methodology.

I reject the idea that it is and argue that, properly positioned, in relation to creation and its metaphysical structures, it can elicit characteristic and appropriate responses to the presence of personal life and agency.

I attempt a number of related things in what follows. All of these are underpinned by the notion that arguments for and against God's existence that try to provide deductive proof cannot in principle be resolved by finding incontestable proofs on the one hand and irrefutable refutations on the other.

Arguments between atheist naturalists and theists often turn on a claim of the experience of the personal in the natural and broader world. Whether someone can be brought to this experience by rational proof alone is unlikely.

The central pivot on which this turns is the observation that the sentient personhood of another is not dependent on first establishing unassailable criteria of identification. The personhood of another is established in mutual engagement with them. I briefly discuss a model of the engagement of God's personhood in nature by Michael C. Rea[19] and show how my conception of the artifactual conveyance of purpose and meaning that "purely" physical *and* human creations allow can facilitate this engagement.

Chapter 6

Ultimate Answers: Meaningless, Impersonal, or Personal

In this chapter I argue that no conception of a "natural world" that excludes the teleological and the personal from a metaphysical framework could provide an "ultimate" explanation for the fact of anything existing at all or its coming into being. Furthermore, I contend that the even the inclusion of abstract entities, such as those of mathematics or Platonic universals, into that ontology would not be sufficient for that purpose since they are causally inert, atemporal, and not ontogenic.

I compare "impersonal" accounts of ultimacy with "personal" ones and show how recourse to the "personal" fulfils the metaphysical need I have identified to explain the presence of normativity, teleology, and aesthetic properties as base ontological necessities.

I devote a section of this chapter as a consequence to discussing the nature of natural theology, of which my argument is an example when rightly conceived. As might be expected, I suggest that if natural theology begins from the premise of *accepting* the "personal" and "impersonal" metaphysical duality of the world that I have been concerned to undermine, then it faces problems that are predetermined by that position: primarily a commitment to the primacy of the "impersonal" understood as the territory from which it must begin its work.

That position having been implicitly (or explicitly) adopted, certain consequences follow, such as the view that the existence of God is a scientific hypothesis, or the existence of God offers a "better" explanation of the data of the world than other "theories"; this is a confusion resulting from conflating method and metaphysics.

The understanding of the world as ultimately the outcome of divine personhood is not a matter for empirical decision; it is a matter of formulating a view of the world which allows for the experiential emergence of divine personhood in the natural. While one might be guided rationally to that possibility, there can be no deductive proof of it. The acknowledgment of the personhood of another emerges in the engagement of oneself as a

19. Rea, "Divine Presence."

person with others, whether human or divine. It does not follow from prior deductive conviction.

Chapter 1

A Sketch for Re-Enchanting Nature

Science, Normativity, and Metaphysics

SUMMARY

In this chapter I contest the view that the achievements of the modern physical sciences have "disenchanted" nature. The "disenchantment" in question derives from a principle of scientific practice that requires no reference be made to divine or personal agency. Aesthetic properties which express such things as beauty are also excluded from "objective" physical nature because their detection and evaluation involve personal experience and judgment. The world of "physical nature" is thus presumed to be barren of any real feature that might show it to be the outcome of creative intent and value.

I outline an alternative metaphysical conception of the world that unifies the existence of a causally governed natural world with persons, values, and expressive properties. I maintain that the very idea of a "natural" world, or "nature" accessible to scientific practice in its various manifestations, depends on this overarching picture that shows their various relationships to be part of a basic ontological union between them.

OVERVIEW OF THE CHAPTER

The central idea of the following chapter stands in opposition to the framework of a modern understanding of the "natural" as a place of causal events ultimately divorced from teleological concepts like "purpose" and expressive properties like aesthetics.

The metaphysical picture I am concerned to undermine conceives objectively occurring and uniformly repeating physical events as describing the ontological entirety of the cosmos: they answer to no final purposes and no other normative judgments. "Normativity," the judgment of an event, object, or property as meeting or failing to meet some standard exterior to any intrinsic properties of these things themselves, has no fundamental place in this final picture of the natural world.

A variety of views of what ultimately exists derive from this basic schema. Because the methods of the sciences proceed *etsi deus non daretur*, "as if God did not exist," the conclusion of most of these is to argue that the natural sciences, particularly physics, are the supreme arbiters of the basic ontology of "reality."

Other "expressive" properties which might indicate intention or purpose like those of the aesthetic and the ethical, for example, which are open to normative evaluations like "This is ugly and could have been more beautiful," are also excluded. If they have an explanatory or ontological status at all, they are only granted it insofar as they are confined to the interior world of subjective consciousness of persons and their conceptual practices. There are, to put the matter simply, no such objective realities out there in the world of things actually beautiful or hideous, but that our thinking makes it so.

Two distinct worlds evolve from the impasse that consequentially develops between the "human" world and the physical one: the "personal" and the "impersonal." From this distinction emerges metaphysical naturalism: the assignment of ontological priority to the latter.

The world is thus taken to be "disenchanted"; no evidence of "meaning" or "purpose" is to be found among its primary ontological constituents. Along with the discarding of these teleological ideas, the possibility of a designer ultimately responsible for the ultimate shape, functioning, and existence of natural things expressed through their structures is similarly dismissed.

I oppose this view. I suggest it is metaphysically, not empirically, derived and cannot escape requirements to justify itself in metaphysical terms: a paradoxical exercise in normative evaluation. Its metaphysical provenance opens it to metaphysical critique. The criticisms I develop consequently form the basis of an alternative metaphysical understanding of the natural world.

As part of this exercise, I demonstrate that the methods of science themselves assume various normative standards which cannot therefore be explained by any science. Among other things, for example, theories depend on the normative requirements of "evidence," the epistemological bedrock of scientific activity.

All this is in preparation for the following chapter. I give a brief summary of its content here, to orient the reader towards the goal it will aim at. The burden of my view in the later chapter rests on the metaphysical principle that anything which can be said to exist must come in some structural form. Structures, for example, must have parts and wholes, consist of individual elements in some relation to each other, be subject to change, be mathematically describable, and have aesthetically expressive properties accessible to normative judgment. This prepares me for the case that aesthetic judgments can be contributors to the representational truthfulness of any theory in the primary physical sciences. I provide evidence in that chapter that mathematical physics uncovers beauty in the basic structures of the physical world.

PART 1

Ejecting "Final Causes" from Natural Philosophy

In the canon of Western philosophy the most prominent of the classical philosophers, Aristotle and Plato did not make distinctions between understanding nature, teleology, and persons in the manner that is now widespread in Western culture. Mats Wahlberg summarizes the dominant pictures of the Greek philosophers as follows: "The nature of Plato and Aristotle was a meaningful order. Things were seen as embodying forms or ideas and existed in order to express those ideas."[1]

For the Aristotelian, for example, philosophical analysis provided a series of principles that could be deployed to explain the activities of anything in nature (or *physis*) in one unifying doctrine. "Nature," for Aristotle, consisted in everything sublunar—that is, below the stars. The superlunary realm of the starry world was fixed and perfect. The events, processes, and objects of nature in the "natural world," however, were in constant change and flux. Each kind of thing—a horse, a flower, a tree—could be understood in virtue of four underlying principles or "causes" that contributed to understanding what it was, what made it the kind of thing it was, and how and why it behaved the way it did.

He considered there to be four primary causes whose interactions provided the required explanations: the *material* cause ("that out of which"

1. Wahlberg, *Reshaping Natural Theology*, 53.

something is constructed); the *formal* cause (the "essence" of a thing, or its primary principle of organization); the *efficient* cause (that responsible for the "motion" or change of an object); and the *final* cause (the end, goal, or purpose of the object, or what it is good for). As well as their application to the animal, plant, and human worlds, these principles of change and becoming could be applied to human artifacts, with the significant proviso that the "essence" of an artifact was imposed on it from the outside. In itself, it had no principle or end to which it spontaneously moved.

Aristotle's four causes

Aristotle's rational principles bound all natural things together into different forms of organized "matter," each kind with its own "essence" and purposes. The behavior and nature of inanimate objects and sentient life, including ourselves, the "rational animal," could be captured in terms of these four causes. Thus, the purpose or final cause of an oak seed was to become a tree, and all its potentials in this respect worked together with the appropriate causes to bring this end about. This understanding of the natural world, which survived well into what we call the "medieval world," was slowly replaced by another, one which replaced the Aristotelian schema with the idea of the cosmos and nature as operating like machines. This development was roughly concurrent with what is now known as the Scientific Revolution, beginning around the late sixteenth century.[2]

2. Much modern scholarship traces the emergence of modern science to multiple philosophical and theological strands extending back into earlier periods. However,

In an arresting image which he exploits in his book *Science and Spirituality: Making Room for Faith in the Age of Science*, Michael Ruse,[3] a philosopher of science and a naturalist with a strong interest in the possibility of religious belief, divides the ancient Greek and contemporary pictures of the world into the difference between the world as "organism" and the world as "machine."

The point of Ruse's choice of the word "organism" is to stress the picture it evokes of "vitality": creatures and living things of all kinds moving and evolving towards a structural end which embodies their final purpose and flourishing, the "essence" of the particular of the kind to which they belong. The word "mechanism" intimates a different world—one emptied of the vitality of intrinsic purpose seeking creatures and life forms bound together by a rationally organized natural world. In its place, the idea of a "mechanism" evolves, one which seeks to explain all natural events, those of the cosmos and those of creatures as units with parts bound by nothing other than a concatenation of causal parts operating meaninglessly together. Peter Harrison explains the change:

> In the sphere of natural philosophy, the Aristotelian understanding of qualities, essences and ends as intrinsic properties of natural bodies was . . . subjected to searching criticism. Change in the natural world began to be understood in terms of laws of nature that were thought to operate externally on inert particles of nature.[4]

Wahlberg elaborates the growth in the conception of the understanding and explanation of nature that this involved:

> In the 17th century . . . the idea of a meaningful order (in nature) was put under pressure by the rise of the new science and came increasingly to be seen as an anthropomorphic projection onto nature of a meaning and order that are not to be found there. In order to achieve a correct understanding of nature such projections must be avoided. The vision of a meaningful order gave way, eventually, to a view of the world as a "domain of ultimately contingent correlations to be . . . mapped by empirical observation."[5]

I am not writing a historical treatise here, but only tracing an idea that became especially prominent and that remains visible in many contemporary popular, and even academic, discussions.

3. Ruse, *Science and Spirituality.*
4. Harrison, *Territories of Science and Religion*, 85.
5. Wahlberg, *Reshaping Natural Theology*, 53.

Teleology and values are thus to be eliminated from the natural world for good methodological reasons but also for good ontological reasons.

By the end of the eighteenth century the Newtonian picture aligned to the ejection of God from the interior of the universe had triumphed. Alexander Koyré expresses the victory trenchantly:

> The Newtonian God reigned supreme in the infinite void of absolute space in which the force of universal attraction linked together the atomically structured bodies of the immense universe and made them move around in accordance with strict mathematical laws. Isaac Newton's three laws of motion, their simplicity, range of explanation, absorption of the Keplerian laws of motion of the planets, and the new mathematics of calculus delivered an opening of the universe to mathematics of extraordinary descriptive and predictive power.[6]

Despite Newton's diffidence in offering a physical explanation for gravity, his own theological attachments, and the suspicion of some of his contemporaries that "gravity" in his conception of the forces of the world was "occult," his work yielded a cosmos contained within a framework of absolute time and space that operated with a mechanical certainty. This picture helped open one way to an understanding of how human thinking and agency came to be thought of in relation to the natural world, man, and God, nicely captured by E. A. Burtt in his classic *The Metaphysical Foundations of Modern Physical Science* when he describes the separation of man from nature and consequently from God as a consequence of the "modern conception" of nature:

> Instead of causal explanation in terms not unsuited to a metaphysics which regarded man as a determinative part of nature, and a link between man and God, we now, after his banishment from the real world, explain causality solely in terms of forces revealing themselves in the mathematically expressible motions of matter itself.[7]

Wahlberg nicely captures the picture that resulted from the extrusion of teleology and the personal from the causal mechanics of the time.

> The process of disenchantment . . . is a . . . cleansing of our world picture from anthropomorphic projections. When everything tainted with human subjectivity is removed from the picture, what is left is the world as it is in itself. . . . Reason, intelligible

6. Koyré, *From the Closed World*, 274.
7. Burtt, *Metaphysical Foundations*, 92–93.

structure, logos, and value belong on the subject side of the subject/object divide.[8]

Modernism thus achieves a decisive dichotomy, making it possible to *conceive* of a mechanistic world[9] which requires at no point in its conception, sustenance, or explanation reference to the teleological, let alone the personal, and, at an even further distance, the divine. "Science" has made the progress it has made from that time by continuing to exclude the operations of all human *and* supernatural rationality and personhood from the underpinnings of *all* nature. The totality of natural reality is henceforth to be conceived as falling within the complete explanatory scope of the new metaphysical picture of reality. The Kantian scholar Robert Hanna expresses the principle in the following way: "The *entire* world including living organisms, animals, human beings, and persons—operates according to non-teleological, mathematic-physical principles alone."[10]

I take it that this description, which is not Hanna's own conviction, is intended to admit into the pantheon of ontological properties essential to the world's constitution only those which are nonphenomenal, impersonal, and perspectiveless: expressive properties like those of the aesthetic which clearly involve phenomenal experience for their appreciation, satisfy none of these criteria, and are excluded from the basic ontology of the universe.

Reasons for Skepticism About Re-Enchantment

From this perspective, to try and re-enchant the natural, as I want to do, is to continue to insist that the physical world external to us manifests primary properties such as the aesthetic and thus be compelled to re-inject normativity into nature. The first immediate objection that arises to this is the thought that if the world external to us did express such qualities as goodness and beauty as part of its physical makeup, then the detection and evaluation of those properties would surely feature in at least one basic science of the physical. The burden to provide evidence for any such assertion would be placed on the protagonist of any such view.

Another objection arises from a necessary feature of the *methodology* of the natural sciences: even if aesthetic properties and other evaluative

8. Wahlberg, *Reshaping Natural Theology*, 54.

9. The word "mechanism," as I intend it to be understood here, is not restricted to the technical idea of a "machine" but also embraces the broader conception of any natural substance or property whose function can be understood without reference to "spiritual," "vital," or "personal" forces.

10. Hanna, *Kant, Science, and Human Nature*, 8.

dimensions *were* present in the natural world, this would anyhow close them off to scientific investigation because they would not be open to objective measurement, the fundamental basis of scientific method. As physicist Jane McDonnell succinctly expresses it, "Scientific theories are supposed to be objective, and beauty is a subjective criterion."[11]

To judge the physical events of the natural world according to some external norm provided by, say, aesthetics or ethics might imply that their causal functioning is governed by some external aim or "final purpose" towards which they are directed and succeed or fail to attain. To introduce aesthetics into the practices and judgments of the sciences, it is then contended, threatens a return to a world in which animated forces, spirits, magic, and arrays of gods are invoked to explain natural phenomena.

In the world of causal regularities, there is no such conception: "failure" is simply the outcome of a complex interrelationship of causation with outcome *x* rather than usual outcome *y*. To say *y* "ought" (indicating a "normative" standard) to have happened is, in the scientific context, just to say that the normal outcome of a set of events has been interrupted by a causal event that is not a regular member of the usual set of causal events constituting the sequence. The common use of "ought" in this context is simply a useful shorthand heuristic, not an indication of a normative judgment. Or so the argument goes.

There is a further objection which develops out of the idea of something being designed. Normative properties are intimately linked with the concept of design. It is where they have their natural conceptual home. "Design" suggests a working towards a particular end (a "final cause"), the achievement or non-attainment of which can be judged against how well it achieves that end. The parts of a designed artifact also owe their structures to the contribution they make to that final purpose. Of course the conception and development of the artifact will require that those parts fit together in ways that are described as "harmonious," and so on. Structural predicates like "minimal," "elegant," "harmonious," and a host of others enter into both how the end is conceived and how it is developed. Throughout the process aesthetic judgments help determine how the end is to be achieved, and on the completion of the artifact, aesthetic judgments of the whole will be applicable. Quite commonly, the idea that a piece of equipment is "awkward" or "ungainly" or "too complicated" is a reason for rejecting or modifying the object or process in question. Ethics, of course, also invokes patterns of behavior and thought directed to ends towards which we ought to aim, and in virtue of which our acts are evaluated.

11. McDonnell, *Pythagorean World*, 16.

In this sense, both aesthetics and ethics participate towards maximizing and giving value to that which they govern. Nothing in nature, however, for the materialist and naturalist is "designed" in the requisite sense. A whole raft of possible intentional judgments are ruled out of any serious explanation of physical events—for example, that "the eye was purposefully designed for sight," "the structures of mathematical physics manifest intended beauty," and "the structures of physics were rationally constructed to be understood." Such explanatory propositions, in views like this, can have no purchase on understanding natural events other than as recognized anthropomorphic shorthands for causal descriptions and functional observations.

Ontological conclusions can quickly flow from the exclusion of normative propositions applied to the natural world. If nothing in nature is the proper object of such judgments, then surely aesthetic properties like beauty, or those features of things that appear to signal design or intention, do not belong to the items in question but are functions of our own psychology; I ascribe "beauty" to the flower not because it is in fact so, or, perhaps, was intended to be so, but because it pleases me in a certain kind of way. "Beauty," in other words, is not an "objective" fact in the world but belongs inside the head. It is not too difficult to understand the concern of the scientist here. She wants to keep her discipline tightly objective and continuing to be truly revelatory of the inner causal workings of nature, and who can deny the power of the achievements of the natural sciences in this regard? We turn to this problem in the next section.

The Aesthetics of Nature: "Beauty" Naturalized?

Despite the cautions and objections that issue from the understanding of science and ontology that I have sketched, we nevertheless find it difficult to separate out many of our phenomenal experiences of the natural world from our scientific and theoretical understanding of natural events. Aesthetic reactions to the external world are common to our day-to-day perception. Even scientists across a range of disciplines are often unable to restrain aesthetic responses to their findings outside the strictly formal context of academic communication.

The physicalist naturalist is left with an explanatory burden as a consequence: how to account for the aesthetic effects of the natural world on us in all its manifestations and the place aesthetics plays in human lives and creations. The intuitive modern answer to this is that if aesthetic properties are not extant in a universe absent of any sentient life, then they must belong to, or be somehow causally dependent on, the internal experience of appropriately equipped organisms like ourselves.[12]

"Naturalist" explanations of aesthetic experience are consequently sought in the life of human persons in the domains of sensory awareness, cognition, perception, and affect as investigated by the psychological and cognitive neurosciences. The task of these disciplines is understood as twofold: to explain the *ontology* of aesthetic properties as an outcome of human neuropsychological structures, and *how* they are generated by the "information processing" of cognitive and affective mechanisms.[13]

Why these mechanisms develop and why they are important to human life are often argued to be discerned through understanding evolutionary processes in which aesthetic experience is tied to survival or an epiphenomenal *spandrel* of some such outcome.[14]

One might call this a modified "naturalized" Kantianism: while there might be assignable aesthetic properties of things that emerge from how we categorize and experience the objects and properties of the world about us, they reveal nothing of the things themselves.[15] Common expressions of aesthetic pleasure occasioned by natural beauty are explained not by features of the aesthetic properties of things themselves but by various evolved cognitive mechanisms serving other psychological purposes: they might be associated with features of the environment that serve the ends of survival—the sparkle of sun on water, for example—and thus provide a useful cognitive heuristic for categorizing very different useful natural items.

From this perspective, even universal agreement about what is beautiful and what is not would not support the idea that aesthetic properties are intrinsic to natural things. It might be thought that if there was agreement across cultures and times about what was truly beautiful or not,[16] this would be sufficient to establish the possibility that beauty was on objective feature of things. But this is not logically the case for the simple reason that everyone could be wrong. Agreement is not sufficient to establish truth. Even if it were the case, which it is clearly not, that every person *was* agreed on what

12. Budd, *Aesthetic Appreciation of Nature*, is an example of a sophisticated full-length philosophical treatment of an understanding of aesthetics in nature that supports this view.

13. Currie et al., *Aesthetics*, contains a series of essays debating the relative contributions of the cognitive sciences to questions in aesthetics.

14. I discuss this in detail in a later chapter.

15. Crowther, *Kantian Aesthetic*.

16. It may well be the case that there is such agreement. Establishing this is an empirical matter. I cannot discuss the literature here, and in any case my overall argument does not turn on it, pleasing though the possibility may be.

was beautiful and what not, from the evolutionary point of view that regards aesthetics as purely dependent on the "subjective," this would *still* not be evidence for the objectivity of beauty. Rather an evolutionary explanation would be developed to account for the commonality of subjective agreement that located the perception of aesthetics entirely within the species.

From within the omniscient resources of psycho-evolutionary explanation, should all human cultures and persons agree that "*x* is beautiful," it would "be true" not because *x* is indeed beautiful as a matter of fact but because, for some arcane sociobiological reason, we had evolved thus. Aesthetics on this view still has its ontological home in the interior subjective experience of sentient persons and not in the impersonal world of the objective. On this view we would simply have all evolved to find the same kinds of things "beautiful."

The division between the "subjective" and the "objective" also has consequences for how the artifacts and aesthetics of human activity are to be understood. Without the human social life surrounding it, no object or artifact that is an outcome of human agency is supposed to have aesthetic value or any other expressive property as a matter of its *intrinsic* constructive properties. Its ultimate identity is simply a concatenation of physical elements aggregated together. There are complicated reasons to be mined from history, social psychology, and a raft of other disciplines related to human life that might fully explain the identities and aesthetics of human constructs in this view.

For this reason, and the importance that music, art, literature, buildings, and other artifacts play in human life, theories of aesthetics[17] are largely concerned with the objects of human creation. Here, it is straightforward to see, created objects are intended by their authors to express certain meanings. Familiarity with the author, painter, builder, or architect and their culture, times, and traditions allow us to adjudicate the aesthetic properties of a piece of work. The symbolic meaning of such works and their aesthetic appeal, however these are to be understood, cannot, so it is thought, derive from the purely physical properties of the object; these are the preserve of the physical sciences and their methods alone. So we see the glimmering origins of a metaphysical development that I discuss in forthcoming sections—a division between what is owed to human intentionality and what is owed the causal processes of the physical. The former is to be

17. Taliaferro, *Aesthetics*, is a good introduction to contemporary philosophical ideas in aesthetics. An advanced discussion of aesthetic judgments, principles, and aesthetic properties is found in Budd, *Aesthetic Essays*. See also Levinson, *Oxford Handbook of Aesthetics*.

metaphysically located in the inner world of human consciousness and the latter in the natural world exterior to it.

If recognizing the hand of the divine in nature requires the detection of personal qualities, or properties like beauty expressed in natural structures, then where are these to be sought in terms of causal forces revealing themselves in the "mathematically expressible motions of matter," it may be justifiably inquired.[18] Nature can be described and explained as doing what it does without any such teleological commitment to anything associated with the cognitive and affective life of human persons, let alone a divine personage.

In some philosophical accounts, even cognitive and affective life are anyway reducible to some materialist account in neurobiological terms which will extract from them those features such as the intentionality of rational consciousness that appear, but only appear, to place them outside the realm of the physical. "Nature" here has an expansive referent intended to exhaust the idea of what the world ultimately is, and contains. Undoubtedly, this is both a powerful argument and a well-established picture of the nature of reality in the contemporary world. In the next sections I begin an attempt to unravel the foundations of this naturalist picture of aesthetics and nature.

Ontological Naturalism and Dualistic Metaphysics

And so the sundering of the teleological, the normative, and the personal, and all the paraphernalia of subjective consciousness, from the natural world and *its* explanatory forces results in two *distinct* worlds, which manage to operate in tandem but whose ontological and epistemological links remain mysterious if traversable at all.[19] Robert Hanna sums this up precisely:

> The fundamental problem . . . is how to reconcile two . . . seemingly incommensurable metaphysical conceptions or "pictures" of the world. One the one hand the non-phenomenal mechanistic, objective, value neutral and amoral metaphysical picture of the world delivered by pure mathematics and the fundamental

18. Burtt, *Metaphysical Foundations*, 92–93.

19. The classic statement of this position was published by the philosopher Wilfred Sellars (1962). The essay, "Philosophy and the Scientific Image of Man," first appeared in *Frontiers of Science and Philosophy*, the first volume of the University of Pittsburgh Series in the Philosophy of Science. It was then published in 1963 as an essay in Sellars's book *Science, Perception and Reality*; it has recently been republished (Ridgeview, 2017). An issue of *HumanaMente* fifty years later devoted an edition to a series of articles on current perceptions of Sellars's position and its history: see Gabbani, "Between Two Images."

> natural sciences . . . on the other . . . there is the subjective, phenomenal, teleological, value laden, person-oriented, and moral metaphysical picture . . . yielded by the conscious experience of rational human beings.[20]

From this separation arises philosophically fertile dislocations. Human consciousness and its concerns are radically sheared from the physical world at the same time as humanity is incarnate in that world. The mind–body problem is the philosophical expression of the disjunction,[21] as is the apparent chasm bequeathed to us by David Hume, the "fact–value" dichotomy, or the problem of normativity in the natural world.[22] The debate about where to locate moral ontology, if anywhere at all, is another example of this division, as is the consignment of aesthetic ontology to the interior of the subjective mind, or the restriction of ultimate moral authority to its emergence from the community of *Homo sapiens.*

The overall picture of the world that this presents us with presses the fundamental question I introduced above: Where are consciousness, subjectivity, teleology, moral ontology, aesthetic appreciation, reason, thought, meaning in language, and all that go along with them to find their place in the broader scheme of reality?[23]

---

20. Hanna, *Kant, Science, and Human Nature*, 8.

21. It is worth noting that for the ancients the concept of "consciousness" as a unique "stuff" would not have been comprehensible. Aristotle, for example, did not suppose such "stuff" to be the defining feature of humanity. His definition of the distinguishing feature of man as a "rational animal" locates *Homo sapiens* as an inhabitant of the natural world describable within his four causes. For a wide-ranging discussion of different Greek models of mind, see Long, *Greek Models.*

In a fascinating discussion, the German classicist Bruno Snell points out that the idea of a unified body was not obvious to preclassical Greeks, who thought instead in terms of different bodily parts. So even the concept of a unified human corporeality is itself an achievement of human thought. See Snell, *Discovery of the Mind.*

22. See Morris and Brown, "David Hume." See also Millican, "Hume's Fork." Millican argues that the discussion of Hume's fork in Hume's "Enquiries" is a simplified one. The original distinction, he argues, is treated more substantially in Hume's earlier work, the *Treatise*, but is nevertheless a poor reworking of Locke's theory of relations. The most obvious objection raised against Hume's distinction is to ask on which side of his "fork" the distinction itself falls. It is neither a deductive truth nor one derived from sensory exposure to the world, the only two forms of legitimate truth claim Hume allows. Why then should it itself count as an acceptable principle governing what there is and what is to count as knowledge? For a wide-ranging discussion of the "fact–value" distinction that supports my contention, see Putnam, *Fact/Value Dichotomy*. Putnam makes the valuable point that a distinction is not necessarily a dichotomy

23. Baker, *Saving Belief*, argues against a wide range of physicalist conceptions according to which properties of mind fundamental to human personhood and mindedness, including teleological aspects such as "belief" and the intentionality of thought,

If that place is entirely defined by some version of physicalist naturalism that admits into the ultimate ontology of the cosmos only what the physical sciences legitimate, then the apparent features that characterize our existence as persons will be ultimately excluded, reduced, or supervenient on one or other of the items of the naturalist scientific disciplines,[24] and the assuaging of our need for purpose and meaning will only be generated from within our own cognitive resources.

At best, perhaps teleology and all the apparatus of human consciousness and life fit into the physical scheme of reality as emergent, supervenient, or properties of the special human "sciences," which are dispensable when it comes to a final accounting of the ultimate constituents of the universe. In this conception teleology, expressive properties, value, and all the rest are extrusions that emerge with human and animate life. If conscious life were to disappear, then so would worlds of value and meaning.[25]

David Papineau, a prominent philosophical naturalist, is clear that all natural phenomena are physical, and all events are ultimately explicable by the laws of physics.[26] His assertion is reflective of a widely held opinion in philosophy and general intellectual culture. The introduction to a volume titled *How Successful Is Naturalism?* expresses it this way: "Naturalism is not only the most accepted creed among analytic philosophers but a widespread view throughout contemporary culture."[27] The prominent naturalist philosopher of mind Jaegwon Kim opines,

---

are reducible to physical structures. Included among her targets are prominent naturalists such as Jerry Fodor and Daniel Dennett.

24. This is not to say that any of these relations will include all the phenomena. Some may bear a particular type of dependency on different forms of the physical and so on.

25. The book-length case in defense of teleological realism is Sehon, *Teleological Realism*. Sehon defends the idea that we commonly explain human behavior by citing the reason or purpose for it. Such explanations, he argues, are absolutely essential to how we understand human agency: a teleological citation of the goal of any behavior is not a causal antecedent of the behavior in question in any physicalist or naturalist sense of that term.

The relation between the logic of reason in teleology and causality in "nature" is a source of contention in philosophy. A classic paper published in the mid-twentieth century has been the source of much discussion. Donald Davidson argued in this work that primary reasons for action could be described in terms of "belief" or "physical" events, the events in question holding independently of either description; see Davidson, "Actions, Reasons, and Causes." A book published fifty years after the paper contains a series of essays discussing the plausibility of this claim; see D'Oro and Sandis, *Reasons and Causes*.

26. Papineau, *Philosophical Naturalism*. He makes his position clear, both in the introduction and elsewhere, without ambiguity.

27. Gasser, *How Successful Is Naturalism?*, 3. This work contains a wide range of

> If current philosophy can be said to have a philosophical ideology it is unquestionably . . . naturalism. . . . Philosophical . . . naturalism has . . . constrained . . . philosophy as its guiding creed for much of the twentieth century.[28]

Ontological Naturalism as a Deduction from Science

Understanding the relation between the natural sciences, the personal, teleology, and the divine through the lenses of ontological or physicalist naturalism with attendant suspicion of the former three categories does not, however, follow *automatically* from the sciences of nature. A way to approach an alternative conception is to begin by drawing a distinction between scientific methodologies and metaphysics.

Philosophers of science and those interested in relation of the natural disciplines to the divine sometimes distinguish between methodological naturalism (MN) and ontological naturalism (ON). While MN does not need to be formally embraced by practicing scientists, it is intended to indicate that scientific theory and practice is successful when it excludes supernatural forces and teleology.[29]

The exclusion of teleology and supernatural forces from the investigative and theoretical practices of the sciences (MN), however, must be distinguished from a metaphysical interpretation about what such an exclusion might mean for the ontological character of the world.

In the words of Peter Harrison, a former Idreos Professor of Science and Religion at the University of Oxford, ontological naturalism is

> the position that goes beyond a mere methodological stance to assert that in reality there are no supernatural agents or forces, and that natural sciences have the potential to explain everything.[30]

essays on naturalism. The literature in this area is extensive. A recent paper by Elpidorou gives a useful overview of physicalist naturalism, its various forms, and its principal concerns. See Elpidorou, "Character of Physicalism."

28. Kim, "American Origins," 84.

29. Harrison and Roberts, *Science Without God?*, contains many interesting chapters reexamining the history of naturalism. One message to draw from the volume is that the historiography of scientific naturalism is not simply a slow and necessary rejection of God from the natural world as a matter of scientific and philosophical progress. Harrison's opening introduction locates this history within the context of current debates about naturalism and its relation to the modern sciences. It serves as a useful introduction to many issues that I have not been able to address here.

30. Harrison, introduction to Harrison and Roberts, *Science Without God?*, 3.

We should be careful to distinguish, however, between two ideas of what the metaphysical rejection that characterizes ON involves. On the first, supernatural forces and agents might be taken to be mysterious spiritual forces that operate like the forces adduced in physics that bring about movement and change. It is this concern that I mentioned in a previous section that can rightfully motivate a rejection of aesthetic and other value properties as objective features of the natural world. This is in essence a magical property ejected from the physical world by the natural sciences. The second idea relates more closely to the picture of a human creator constructing something novel.

The theist need have no qualms about accepting the first as a valuable epistemological move towards scientific knowledge of the physical forces of the natural. Why the distinction I have made is important is revealed in a comment by Brendan Larvor in his article "Naturalism": "We live in a 'disenchanted,' that is to say, magic free world, not just because we say so, but because our society runs on effective efficient natural causality."[31]

It is clear, I suggest, that he thinks accepting a theistic view of the world commits the believer to the presence of magical forces. He is, however, conflating the distinction between two different conceptions of the relation of God to the physical universe.

The second conception with which I am largely concerned is not vulnerable to Larvor's criticism. The idea of a divine "supernatural" actor in this latter picture is closer to the idea of a person with teleological agency, related to the creation of *all* reality as an agent to artifact. However that idea is to be understood it is not, or should not be, understood as the idea that God as a personal agent is equivalent to an extra force inside the universe as a supplement to the forces identified by physics.

In more developed philosophical and theological ideas, the coming into to existence of the entire cosmos in its intricate causal and physical functioning is the outcome of a divine creative agency. In a traditional theistic tradition, for example, God is the primary cause of the existence of the cosmos in all its physical forms, shapes, and rational intelligibility, and the cause of that in virtue of which they continue to exist and function as secondary causes. Thomas Aquinas is an exponent of this conception.

The ontological inference from MN to ON that Harrison discusses has the effect of excluding the second conception of what constitutes "reality" as well as the first. It has another consequence, however, that of also excluding what seems, at least to human beings, an indispensable set of interlinked teleological concepts, such as "purpose," "agency," "rationality," and "value"

31. Larvor, "Naturalism," 51.

as real phenomena of the world as encountered by us in it, and necessary for the identity of anything to count as a self-conscious, reasoning person with teleological agency *in* the world.

The ontological inference also has the curiously unintended effect of excluding any definition of scientific activity as "the purpose of gaining knowledge about the physical world." Since epistemological purposes are teleological, the definition assumes that the "purpose of gaining knowledge" expresses a real phenomenon and legitimately explains why persons engage in scientific activity. This alone should provide reason for metaphysical caution.

To question the metaphysics of ontological or physicalist naturalism is to inquire whether an alternative metaphysical framework can be derived into which the personal, teleology, normativity, science, and the natural world can fit as ontologically essential to any conception of a natural world in which physical objects appear and can be understood through empiric-causal lenses—in other words, without any ontological conflict between the personal and the impersonal worlds.

Ontological naturalism, despite its claim to *reflect* a metaphysical reality deriving from the hard natural sciences, might, if such a picture is available, be seen to *leave out* something about that reality. It may even be that what it leaves out is necessary for the possibility of science itself, the characterization of what sort of thing the world would have to be, to be suitable for empirical investigation.

To choose one of many examples of a metaphysical necessity: In order to underpin what is meant by a law of nature, the conceptual link between physical regularity and causal relation must first be grasped. Since this is a prior requirement of understanding physical events, it imposes a normative requirement on the possibility of the idea of a scientifically investigatable nature.

Metaphysics, Ontological Naturalism, and Science

An initial response might be to deny the coherence of any alternative metaphysical picture whose claims about ontological ultimacy are not answerable to physics, or some collection of the natural sciences. Physicist and priest John Polkinghorne, in his introduction to a book concerned with mathematics and meaning, has this to say in response to that claim:

> Anyone who outlines a world view circumscribing the scope of reality, whether it is narrowly or capaciously conceived is

> making a metaphysical assertion as surely as they are using prose to make their assertions.[32]

In other words, the thought that claims about ontological ultimacy can only be answered by the sciences is itself not a scientific claim but a metaphysical one. Metaphysics, as Polkinghorne's comment implies, is an attempt to attain the "true ultimate explanation of the most general features of the world that we inhabit."[33]

At least, as Timothy O'Connor points out, this has been its traditional role. It is the rational effort to attain a picture of what *kinds* of things—such as substances, events, properties, and even kinds of explanation and relations—ultimately exist *in* and of the world insofar as we can intellectually know it. Metaphysics thus conceived *includes* the sciences as part of its subject matter.

A Christian theologian and philosopher can also accept this description. Andrew Davison's 2019 book titled *Participation in God: A Study in Christian Doctrine and Metaphysics* offers early on a definition of metaphysics with which even the nontheistically inclined might willingly concur: "The business of metaphysics," he proposes, "is to offer an account of the nature of the world, and especially of its fundamental structure at the level of being."[34]

Yet Another Christian philosopher, Keith Ward, uses the word to mean "a general account of the nature of reality and the place of human beings within it."[35]

It is probably an intuitive starting point at least in a Western tradition to include in these definitions the aim that the desired description of the ultimate nature of being will unify all the phenomena of the world within its explanatory scope.[36]

Many authors—philosophers and others—have pointed to the fact that any appeal to the epistemological and metaphysical hegemony of the natural sciences is itself a metaphysical position. This is a position often characterized as "scientism."[37] As R. N. Williams, among others, has of-

32. Polkinghorne, "Mathematical Reality," 28.

33. O'Connor, *Theism and Ultimate Explanation*, ix.

34. Davison, *Participation in God*, 9.

35. Ward, *Sharing in the Divine Nature*, 2.

36. A locus classicus for an opposing view is Dupré, *Disorder of Things*. Dupré argues that the state of the different scientific disciplines supports the view that a unitary science is out of the question, and he develops a metaphysics to accompany this position.

37. De Ridder et al., *Scientism*, presents a comprehensive series of essays both for and against "scientism."

ten pointed out, this is itself a metaphysical claim requiring a naturalist metaphysics for its defense.[38]

Taliaferro et al. similarly suggest this is so for the following reason:

> The claim that the physical or natural sciences should be our primary starting point is itself a philosophical claim. . . . If we claim that these ways of knowing reality are the only or best methods for doing so, we are making claims about physics, chemistry and biology from a perspective external to those disciplines. This is a philosophical claim about science not a scientific claim about science.[39]

As E. Feser argues, "scientistic propositions" are refuted the moment they are stated because they are themselves "metaphysical positions that could only be justified using metaphysical arguments."[40]

The collective weight of these considerations supports the role of the need for a perspective outside the sciences that can participate in the quest to outline "the most general features of reality." It admits requirements for *general features* of the world that are not encompassed by the sciences but are necessary for them.

The Metaphysical Idea of a "Scientific" Nature

One result of all these observations is that the idea of "nature" or different "natures"[41] as describing a world external to us that can be understood scientifically is an outcome of a particular general way of understanding the world. It is not a *product* of "science" but rather a *precondition* of it.[42] The harmony between scientific practice and world that is epistemically productive now seems to our scientific culture a "natural" way of comprehending

38. Williams and Robinson, *Scientism*, offers a series of essays covering a comprehensive range of issues surrounding epistemology, metaphysics, theology, and the humanities in relation to naturalism.

39. Taliaferro and Meister, *Contemporary Philosophical Theology*, 22.

40. Feser, *Last Superstition*, 84.

41. Snell and McGuire, *Concepts of Nature*, as the title suggests, offers a historical overview of how human beings have understood, or come to a conception of, "nature." A particular issue with which the volume is concerned is how the modern conception of a "causal" nature has extruded the idea of an order of "natural value," in virtue of which human beings could seek not only a moral identity but normative guides to action.

42. Heller, "Science and Transcendence."

nature. The harmony, however, is not an a priori given of human sense and cognition.[43]

The characterization of an aspect of a part of world as "nature,"[44] a realm of nonpersonal yet intelligible and structured uniform events, is a *basic* prerequisite of contemporary science.[45] Understood in this way "nature"[46] is a philosophical achievement, the outcome of a confluence of philosophical change, technical and mathematical development.[47]

In her extensive philosophical investigation of an "expansive" naturalism (which I discuss below) Fiona Ellis, puts it this way:

43. Brooke and Cantor, *Reconstructing Nature.*

44. I take it here that, despite the difficulties in defining "nature," there is an immediate sense in which the term functions superordinately to distinguish areas of inquiry concerned with those aspects of the world considered, in some sense, "external" to us. Of course, "we" are corporeal creatures, and to the extent that disciplines such as biology and the neurosciences concern human beings, the distinction becomes less clear. Even so, there remains an important difference between such disciplines and fields such as moral philosophy, literature, history, and poetry. A distinction, however, need not imply a dichotomy.

45. It is sometimes supposed that the possibility of a "chaotic" world would undermine this position. But a chaotic world would still require some structure in virtue of which it is chaotic rather than simply unintelligible. The central issues here concern causation and determinism, since "chaos" itself possesses describable causal patterns. Indeed, the very intelligibility of the concept of "chaos" depends upon its contrast with "ordered" structures. As is now well-known, simple mathematical equations, recursively applied, can generate results that are chaotic in appearance while nevertheless remaining mathematically describable. "Chaos," therefore, is not the absence of structure but one mode of structured being.

46. Gal, *Origins of Modern Science*, is a recent contribution to the vast literature on the development of science. Wagner and Briggs, *Penultimate Curiosity*, offers an empirical and intellectual history arguing that the development of scientific theory, method, and reflection on nature was originally animated by what might be called transcendental concerns about understanding the world in which human beings find themselves.

47. As one example, the development of calculus by Newton and Leibniz made possible, for the first time, the systematic mathematical description of continuous change, motion, rates of variation, and the calculation of areas under curves. A further example is provided by the Oxford philosopher Anthony Quinton, who argued that one of the conditions underlying the development of modern science was the theistic conception of "nature" as rationally ordered by a divine mind. Because human beings were themselves understood to participate in that rationality, they were thought capable of penetrating and explaining the causal structure of the world. Quinton further argues that science, in the modern sense, failed to develop within the otherwise rich intellectual traditions of other cultures, such as the ancient Chinese, because its metaphysical conception of the natural world did not include the idea of a rationally structured interior. See Quinton, *From Wodehouse to Wittgenstein.*

> Science provides a framework for learning the rules by which nature operates. . . . [Nature is seen] as . . . events . . . which belong to the realm of 'natural-scientific intelligibility,' and one of the paradigmatic ways in which natural science makes things intelligible is by subsuming them under natural laws.[48]

Nicholas Maxwell describes beautifully and with a simple clarity how we get to the position that Fiona Ellis articulates. I quote him extensively since he clearly describes the movement from metaphysics to methodology, a position which forms the general backbone of my argument as it develops to critique naturalism.

> If science is to be possible at all, some kind of assumption must be made about that of which we are most ignorant: the ultimate nature of the universe. . . . *It is above all here that it is vital that we make as good an assumption as possible, one which does the best justice to the nature of things. For it is this basic cosmological assumption that will determine our methodology, what kinds of theory we are prepared to consider.*
>
> If, for example we believe a society of gods governs the natural world according to whim, then in seeking to improve knowledge and control over natural processes, then it will be entirely rational (relative to this belief) to adopt such methods as . . . sacrifice, oracles, omens (and so on). If we believe a pattern of physical law governs natural processes, quite different methods will be adopted, namely those of putting forward precise hypotheses concerning possible laws governing phenomena to be tested against observation and experiment. Metaphysics determines methodology. . . . A bad basic metaphysical conjecture, hopelessly at odds with the basic nature of the universe, will lead to the adoption of an entirely inappropriate set of methods.[49]

For theoretical science *and* the appropriate methodology to be possible, we need to carve metaphysics at the right joints. We do not just perceptually encounter "nature" and thereby accept the causal independence of nature from the personal and the animate and the possibility of an experimental and mathematical science. We must grasp the conceptual organization of reality appropriately to begin to question the natural world so that it yields up the mechanisms of its causal and mathematical structures.

Maxwell makes that clear, and it is evident from anthropology, cultural history, the multiple varying conceptions of "nature" in different societies at

48. Ellis, *God, Value, and Nature*, 77.

49. Maxwell, *Comprehensibility of the Universe*, 4–5; see also p. 6.

different times, and the history of the development of the natural sciences over centuries. Without the *idea* of impersonal forces and events constituting intelligible physical structures to be intellectually unearthed from the perceptual appearances present to our senses, there could be no science.

A commitment to the appropriate concept of nature and the methods that flow from it provision us with an intellectual framework to which we ought to be committed if we want scientific progress. The words "ought" and "commitment" should not escape our attention. They indicate a recognition that the world of nature conceived as a contingent realm of independent forces rationally penetrable, mathematically and empirically investigatable, is fundamental to any scientific theory, and involves a normative commitment to a particular way of seeing the world external to us.

Discussing the application of mathematics to the natural world, for example, the particle physicist S. M. Barr writes,

> The scientific revolution, as well as the modern science to which it gave birth, was characterised not only by a happy marriage of mathematics and experiment but also a different way of looking at mathematics and its application to the world.[50]

Prior to this advance, mathematics was understood as usefully descriptive but giving no real insight into phenomena and their underlying causes; that was the job of "natural philosophy."[51] So, even the application of mathematics as a productive tool in explaining natural events is an achievement of the intellect in coming to see that the "natural world" is structured in such a way that those structures obey certain principles of logic-mathematical organization.[52] This is an accomplishment of some note, and one which bears much on my thesis in this book.

50. Barr, *Guide to Natural Science*, 37.

51. Barr, *Guide to Natural Science*, 37.

52. See Collingwood, *Idea of Nature*. Although Collingwood's book is now dated, it nevertheless provides an interesting narrative of the development of the "idea of nature" through three major phases of Western philosophy: Greek conceptions from Thales and the early preclassical thinkers through Plato and Aristotle; the Renaissance and early modern periods, including discussions of anti-Aristotelianism, Bacon, Kepler, Spinoza, Newton, and the development of the idea of "matter"; and finally Berkeley's idealism, Kant, and Hegel. His concluding section on the "modern view of nature" is inevitably outdated, but it still serves as an illuminating historical perspective in its own right.

PART 2

Normativity and Science: A Critique of Naturalism

Any attempt to introduce normative properties, aesthetics, and matters of value back into the ontology of the natural world must begin by examining ontological naturalism. As I have discussed, the central proposition defended by most naturalists holds that the only warranted knowledge of ontology is that adjudicated by the sciences.

The epistemological assumption most commonly advanced is the view that science has all the answers to the origins and nature of the universe, *including* those features of it that can account for its own success. Whatever way the natural world is to be understood, the methods of the natural sciences are the only means to warrant *any* truth claims about ultimate reality.[53] De Caro and Macarthur explain this as follows: "The methodological or epistemological scientific naturalist holds that it is only by following the methods of the natural sciences—or, at a minimum, the empirical methods of a posteriori enquiry—that one arrives at genuine knowledge."[54]

It is, as Lydia Jaeger argues, "the disposition that seeks an explanation of the natural order within science itself."[55]

A closer look reveals the metaphysical implausibility of the thought that any explanation of the success of the methods of science *can* be derived from the sciences themselves. If the success of those methods implies prior conditions, then the limiting of reality to what can be revealed by those methods alone will be insufficient to encompass the totality of that reality.

Some naturalists have in fact recognized or accepted that recourse to normative standards is a necessary component in developing anything like a comprehensive grasp of the place of science in the world.[56] They have not, however, expanded that comprehension to include any idea that nature and its intelligible contents may exist, be intelligible, and display normative properties as a consequence of intentional creation. I discuss this more "liberal" or "expansive" naturalism further on.

I begin a closer critique of the central obligation of naturalism to empirical science by examining the position of Gregory Dawes, a naturalist

53. Almeder, *Harmless Naturalism*, provides a good summary of varieties of naturalism while defending a moderate version.

54. De Caro and Macarthur, *Naturalism in Question*, 7. Onyango Okello, *History and Critique*, offers a comprehensive account of varieties of methodological naturalism.

55. Jaeger, *What the Heavens Declare*, 141.

56. Putnam, *Naturalism, Realism, and Normativity*, contains an interesting selection of essays by the Harvard logician and philosopher of science on these issues.

philosopher who has a particular interest in theist claims. He is of interest because he attempts to incorporate "supernaturalism" and the existence of God *into* scientific methodology, asserting that the proposition that "God exists" should be treated like any other explanatory hypothesis available for empirical test. It is one for which he thinks the hypothesis fails.

I criticize his view, exposing his unintentional attachment to normative judgment and aesthetic criteria. His case is outlined in two publications. The first is a book-length treatment titled *Theism and Explanation*.[57] The second is an important paper in the theological literature titled "In Defence of Naturalism."[58] I concentrate on the paper since it effectively summarizes the main argument presented in the book.

My critique of Dawes leads to a consideration of the second form of naturalism, the more "relaxed" or "liberal" version. Under pressure from the sorts of criticisms I offer of Dawes's view, the "liberal" acknowledges necessary cognitive relations between the physical world and persons who bring it within their cognitive and empirical purview. That need, however, cannot be satisfied by any naturalist metaphysics.[59] This is a theme that will run throughout the rest of this book.

Gregory Dawes: Inferring Ontology from Method

In the next few paragraphs, I briefly summarize Dawes's argument to orient the reader to the more substantial analysis that follows. His paper begins by making a preliminary distinction between methodological naturalism and ontological naturalism. The principal motivator of naturalism, he conventionally notes, is to exclude reference to the supernatural and intelligent purposes (teleology). He rightly argues that the physical sciences have achieved much by proceeding as if it were the case that there were no such forces or intervention in the fundamentals of the natural world. This is what I have previously discussed: the principle of methodological naturalism (MN). It is a distinction originally made by the Christian P. de Vries in 1986.[60]

MN entails only that the operations of science may not appeal to supernatural entities and their putative properties in any scientific theory or

57. Dawes, *Theism and Explanation*.

58. Dawes, "In Defence of Naturalism."

59. Goetz and Taliaferro's *Naturalism* calls the first version I am discussing "strict" naturalism. A variety of terms appear in the literature, but they are generally concerned to convey much the same point.

60. De Vries, "Naturalism in the Natural Sciences." Halvorson, "Why Methodological Naturalism?," provides a thorough review of the literature.

explanations of the natural physical world and its processes.[61] The appearance of the divine in any scientific accounts of the natural world is ruled out as a matter of declared procedural principle and has proven productive. It prohibits, for example, an appeal to the "God of the gaps" argument: the argument that tries to find a space for divine intervention in the world where science has so far failed to give an intrinsically causal explanation.

Dawes again rightly maintains that this is a methodological decision that does not commit one to ontological naturalism (ON), the thesis that all that can be ultimately said to exist is the physical, however this latter term is to be understood. He therefore restricts the use of the term "naturalism" to ontological metaphysics. It seems plausible that MN and ON offer a useful distinction rather than a dichotomy that all participants in the debate who are sympathetic to the vast achievements of the natural sciences will be happy to accept.

Because he wants to exclude metaphysical speculation from method, Dawes outlines what he calls a procedural methodology. This makes the sciences dependent on what he calls "publicly available evidence." This, he supposes, has no metaphysical entailments. Nevertheless, he argues that the historical empirical weight of the evidence accumulated by proceeding *etsi Deus non daretur* contributes to a supposition in favor of ON.

I contest his separation of metaphysical involvement from procedure. The general critique of his position is that outlined by Maxwell above, suggesting that methodology will follow from the metaphysical assumptions made about what ultimately constitutes the cosmos. Following Maxwell's line of thought I maintain that the metaphysical features which characterize the ontological presentation of the natural world *also* allow for the procedures and methods of science that investigate them.

Dawes as I hope to show, ironically and unintentionally agrees with the general idea that assumptions must be made about the broader metaphysics within which science must operate when deciding between theories that go far beyond his procedural principle. I demonstrate that far from offering a thesis that provides for a conception of the "divine" or "supernatural" as a scientific hypothesis open to falsification, he in fact opens the question of the place of aesthetics in our understanding of the entities constituting the deep structures of mathematical physics, and the possibility that these properties express a divine signature.

I now turn to Dawes's argument in more detail.

61. Boudry et al., "Intelligent Design Creationism," and Shanks, *God, the Devil, and Darwin*, are among those who have discussed versions of naturalism similar to that defended by Dawes.

Dawes suggests that the MN/ON distinction is normally maladroitly drawn, and that this has malign consequences for the proper understanding of the claim for ontological naturalism. He opines that this leaves it open to the accusation that MN, in fact, *imposes* the metaphysical requirement that natural processes are to be explained only by reference to other such processes *prior* to the science, thus eliminating a priori potential theistic explanations by definitional fiat.

Some authors, he suggests, such as Johnson, think that this is not a coincidence, "for scientific materialists [or physicalists]," he says Johnson maintains, "the materialism comes first; the science comes thereafter. We might more accurately describe them as materialists employing science."[62]

This is a criticism with which the theist has much sympathy because of its acknowledgment that materialism does not emerge from the sciences but is a metaphysical assumption that comes before scientific practice.

In any case, Dawes correctly diagnoses the error. He argues that the MN distinction as it is commonly drawn involves a confusion between a method of inquiry and the metaphysics of naturalism. He proposes that MN is better clarified as an epistemological requirement of scientific procedure.

He raises W. V. Quine's[63] definition as an example of the confusions that arise when the distinction is not made sharply in these terms. Quine is probably the best known and most influential naturalist of the late twentieth century. Quine, he says,

> defines "naturalism" as the renunciation of the idea of a first philosophy, somehow prior to empirical enquiry. It insists that "the only means we have of figuring out what the world is like is our experience of the world and our explanatory theories about it." It follows that if we want to discover what kinds of things exist, we should look to what our best scientific theories are telling us, whatever that may be.[64]

Commenting on Quine's definition, Dawes offers the following justification and description of his own reconstitution of the methodological principle:

> It is unhelpful to describe [the] procedural requirement as a species of naturalism since the term is potentially misleading. *All this procedural requirement demands is that any proposed explanation should be testable against a body of evidence that is*

62. Dawes, "In Defence of Naturalism," 5.

63. Hookway, *Quine*, is an excellent and comprehensive introduction to Quine, containing an outline of Quine's naturalism.

64. Dawes, "In Defence of Naturalism," 8.

> *accessible to any capable observer*. In the natural sciences, this corresponds to the idea that an experiment should be replicable. But the kind of evidence required is not limited to the experimental data typical of the natural sciences. It could also be archaeological or documentary data to which historians customarily appeal.[65]

And he concludes,

> It is therefore a mistake to say that for science to be science, it can by definition pursue, identify, and entertain only natural causes. What we should say is that for science to be science, by definition it "can pursue, identify and entertain" only those causes whose existence can be argued for on the basis of publicly acceptable evidence. This procedural requirement is the *only* non-negotiable commitment of . . . the sciences.[66]

As Dawes admits, and is happy to do, the procedural requirement does not entail an a priori *commitment* to a materialist or physicalist metaphysics. In fact, he suggests that it involves no commitment to *any* metaphysics. Divorcing the procedural requirement from metaphysical naturalism allows him, he thinks, to bring potential theistic explanations of natural phenomena within the scope of investigation that would qualify them as "scientific" theories.

He exemplifies this by reference to the intelligent design theory (ID). Noting a US court judgment against a school board that had sought to include ID into the curriculum, he describes how the presiding judge justified his ruling on the grounds that "ID violates the centuries old ground rules of science by invoking and permitting supernatural causation."[67]

While acknowledging that this hands the opponents of ID an immediate victory, he nevertheless poses the question "What if appeal to a Divine agent *were* the best available explanation of a set of puzzling phenomena?"[68] Crucially, he then claims that there are

> commonly accepted lists of explanatory virtues, such as scope, explanatory power and simplicity, or testability, ontological economy and informativeness. . . . If a range of explanations positing a Divine agent could be shown to exhibit such qualities,

65. Dawes, "In Defence of Naturalism," 8.
66. Dawes, "In Defence of Naturalism," 14.
67. Dawes, "In Defence of Naturalism," 14.
68. Dawes, "In Defence of Naturalism," 13.

> to a greater degree than other proposed explanation, then scientists ought to accept [them].[69]

As I draw attention to further on, the admission of virtues into theoretical decision-making is a crucial one of whose consequences he seems unaware. But having made his distinction, Dawes carries on by noting that most modern scientists (and others including, for example, historians) *do* in fact have a working ontology, which includes a set of assumptions about what is likely to exist. This set of metaphysical assumptions involves a commitment to the view that the functioning of every element of the natural world will be brought under the explanatory purview of purely naturalistic and physicalist assumptions; they will exclude divine agency. He suspects that this is the source of the confusion.

After all, he asks, "what does Methodological Naturalism entail?" and answers: "That we investigate the world *etsi Deus non daretur*: as if there were no supernatural causes."[70]

This leads him to what he supposes *must* follow:

> This means investigating the world as though ontological or metaphysical naturalism were true.[71]

Although the formal outcome of his distinction between procedure and metaphysical naturalism thus yields a potential place for theistic explanation, the devil is in the last line of his quote in the above paragraph. It might be recalled that the title of his paper is "In Defense of Naturalism," and in his last sections he devotes himself to defending the view that ontological naturalism *is* in fact the correct metaphysical stance to adopt.

The details of this need not concern us here, but it involves appeals to the successes of the sciences which have pursued their enterprises based on naturalistic assumptions, and he is confident that no theistic explanations can or will achieve similar predictive or explanatory success. It is hard to forebear noting that this is itself a faithful normative acquiescence to ontological naturalism.

69. Dawes, "In Defence of Naturalism," 13.
70. Dawes, "In Defence of Naturalism," 11.
71. Dawes, "In Defence of Naturalism," 11.

Normativity and "Expansive" Naturalism: The Case for a Non-Physicalist Metaphysics

Let me begin with a simple example. Suppose we have an interest in art. If one then wishes to know something about how oil paints manage to produce the visual effects they do upon our visual system, then certain forms of investigation, the way they reflect light as a consequence of the chemical structures and light waves of the paint concerned, and the reception of that light by aspects of our cortical visual system, involves adopting methods suitable to answer a particular kind of question. This, one might well suppose, would involve examining the oils in question on a given canvas without reference to the painter, to use Dawes phrase, etsi Pictor non daretur (as if the artist did not exist); but it hardly follows from that requirement that one leaps to the judgment that the painter is not likely to exist or have existed, nor does it follow that positing a painter to explain the very existence of that particular artifactual creation should pose any threat to the scientific methods that might be of use in explaining why and how the artist in question used her paint. The painter as an agent is responsible for the arrangement of the paint, the different kind of paints, and so on. This is not in any kind of theoretical conflict with our scientific queries. It answers a categorically different question. So Dawes' target is entirely misconceived; no serious theist or theologically inclined philosopher has ever supposed that God is an object within the universe, nor that his powers are to be hypothetically conceived as forces in opposition to those recognized in the physical sciences. The artist, of course, must adhere to the methods demanded by the material and world to achieve the effects she wants. Analogously, the answer to certain sorts of causal questions about nature involves the bracketing of theistic understanding, and embracing the principles of methodological naturalism, to which we ought to adhere in pursuit of our goals.

Dawes fails to see, in any case, that the demand that scientists *ought*[72] to adhere to methodological naturalism is *itself* a normative demand. He also fails to see that what gives the demand its normative force is an external epistemological standard which cannot plausibly derive from the natural sciences themselves since it is a prerequisite of anything that is to count as a scientific judgment. Wherever these standards are to be found, there must be a broader picture which can account for the explanatory efficacy of MN. It is the normative demands of those standards that the scientist should obey.

72. Kincaid et al., *Value Free Science?*, contains a series of essays on different aspects of value in science. Those most closely related to the issues raised here are contained in part 3 ("Values and General Philosophy of Science Perspectives").

It is at this crucial point that his account fails. To understand why this is so, I look more closely at the idea of "evidence" and normativity. Slagle reminds us of the importance of the normative claim:

> Normativity . . . is "shouldness" or "oughtness." If we should do something, or if we ought not to do it, we are employing a norm, a standard to be met. Normativity's most visible face is ethics, but it has reverberations throughout philosophy: in logic, *aesthetics*, . . . and epistemology.[73]

Let me remind the reader of Dawes's claim:

> It is . . . a mistake to say that for science to be science, it can by definition pursue, identify, and entertain only natural causes. What we *should* say [a normative requirement!] is that for science to be science, by definition it "can pursue, identify and entertain" only those *causes* whose existence can be argued for on the basis of publicly acceptable evidence. This procedural requirement is the *only* non-negotiable commitment of . . . the sciences.[74]

Dawes means to acknowledge here that beliefs, purposes, and final causes might be identified as "causes." He is, I think, forced into this position because, as he says, he wants to include such things as archaeology, history, and so on into the "scientific" disciplines qualifying as they do as evidence-based.[75]

This is a substantial concession. He seems to fail to recognize, however, that the inclusion of teleological concepts into the explanatory ambit of scientific hypotheses raises difficulties for the idea that "causes" in the sense he requires them to be are the ultimate explanatory items. Whether or not a belief functions as a "cause" in the requisite sense—i.e., as something that forces an outcome once present—is, to say the least, a controversial claim when it comes to explanations or making sense of human behavior. It assumes rather than argues for the idea human persons necessarily act on a belief and have no choice whether or not to do so.

The trouble is that most physicalist naturalists don't want beliefs to operate as explanations that escape the determinative functions of physical

73. Slagle, *Epistemological Skyhook*, 96. Slagle's book is the first full-length examination of the argument that naturalism and determinism are necessarily self-defeating. He examines many versions of the argument and discusses numerous protagonists and antagonists, including Kurt Gödel, Karl Popper, and Alvin Plantinga.

74. Dawes, "In Defence of Naturalism," 14; emphasis added.

75. Larvor, "Naturalism," presents the case for the admission of objects of human activity explained through meanings and purposes that fall outside the natural sciences but are nevertheless, he argues, fully explicable within a humanist naturalism.

causes. They want causes whose presence, in some sense of that word, require the outcome to then occur. Teleology, in the way Dawes wants it, allows for teleology, including the possibility of free will in action, belief, and rationality. In effect, all that is excluded in the end is a certain idea of "supernatural forces," acting like physical causes within the universe, and those "mystical" forces whose protagonists believe them to operate like physical causes in explanations, but are not prepared to subject them to the methodologies that can test their claims. Either that or special privileges are claimed for them.

It is contestable, to say the least, that any serious theologian, or philosophically inclined theologian at any rate, understands the relation of God to universe by a strict metaphysical identity of physical cause with divine action.

Dawes is trying to extend the scope of scientific naturalism to include teleology and teleologically based explanations into the realm of scientific hypotheses because he also wants to include the question of divine action *in* the world as a potential hypothesis and not exclude it by fiat—i.e., only by admitting strictly physical hypotheses to begin with. He thinks that by doing this he can avoid the charge that he has leveled at materialists/physicalists: that they themselves *begin* with physicalism rather than coming to it as a consequence of scientific investigation and "publicly acceptable evidence."

So what is evidence anyway? I begin with how Dawes himself entertains the notion. What defines scientific activity he tells us, is this:

> Science can . . . entertain only those *causes* whose existence can be argued for on the basis of publicly acceptable evidence. This procedural requirement is the *only* non-negotiable commitment of . . . the sciences.[76]

"Publicly acceptable evidence" is presumably efficacious in deciding the truth or otherwise of a hypothesis. I take it that this depends on how adequate the evidence is to causally explain a hypothesis, and on this we can certainly agree. That evidence bears on how we decide some hypothesis: "Did A cause B?" "Do As in general cause Bs?" We mostly accept that the investigate procedures of the sciences are the way to set about deciding these questions, and there is certainly a sense in which the acceptance of the answers are communal in how they are decided. But note that any such acceptance of an answer turns on an acceptance of the implicit norm that is embraced. How is that norm itself caused, and why should we obey it? It is, Dawes tells us, nonnegotiable. If it is nonnegotiable, that presumably is because it true without question and cannot be questioned. One wonders

76. Dawes, "In Defence of Naturalism," 14.

from whence it derives this extraordinary normative force. It cannot derive it from "publicly acceptable evidence" because that produces a empty circularity. But it is the whole crux around which Dawes's argument turns.

At the very nexus of this assumption, a problem faces any metaphysical scientistic naturalist, or physicalist naturalist. As Jim Slagle points out,

> The only obvious norm available in naturalism is causality.[77]

This yields a problem of substantial magnitude because if causality is the only epistemic criterion on which a judgment rests, then

> causality is insufficient to function as an epistemic norm . . . [because] false beliefs and fallacious reasoning are "caused" just as much as true beliefs and veridical reasoning.[78]

The force of this criticism lies in the fact that science relies on some conception of truth that involves normative judgment—i.e., we are to appeal to some concept of what is true and what is false to judge that *x* is the case. That the finding of *x* in an experiment disproves our hypothesis is predicated on a normative relation between the "evidence" and the hypothesis.[79]

If judgments about "evidence," whether privately made or before the arena of public judgments, are *themselves* "caused" in the requisite efficient causative sequence, then the possibility of scientific judgment itself collapses; the language of justification in terms of some criteria external to the practice of the science will simply collapse into a causal sequence itself without any normative force. With it goes the very idea of scientific truth as ultimately answerable to the way things are.

Naturalism, at least in its most stringent forms, cannot allow the intrusion of reference to the normative world of truth because it allows only for causal relations.[80] To say that false beliefs and true beliefs are "caused"

77. Slagle, *Epistemological Skyhook*, 96.

78. Slagle, *Epistemological Skyhook*, 96.

79. The philosopher Karl Popper held the view that a scientific hypothesis cannot be conclusively proven true but can be shown false by an experimental result that falsifies it. Discussions of Popper's principle are extensive. A good introduction is Magee, *Philosophy and the Real World*. The original work in which Popper's claim appears is Popper, *Logic of Scientific Discovery*. My assertion above is not intended either to confirm or to refute this view but only to illustrate a more general principle concerning the relation between normativity and evidence.

80. I am simplifying many highly complex positions. Quine, for example, admitted into ontology whatever was required for science to make progress. Both mathematics and predicate logic he regarded as indispensable to the formation of scientific hypotheses, while epistemology in the Quinean scheme is grounded in sensory engagement with the external world. See Hookway, *Quine*. See also Harman and Lepore, *Companion*

in the requisite determinate sense evacuates from them the idea that it is their perceived truth or falsity relative to some criterion of truth that the individual considers before choosing the one that is true. He chooses that, if he does, because "truth" is a value and a standard to which he wants his scientific endeavors to conform.

For a certain class of naturalist, opening the door to cognitive relations between world and mind in this way threatens to introduce the problem of how a world of brute physical events interacts with cognition and rationality so as to produce a conceptually demarcated space within which logical operations can work. In other words, once you have truth, then you must have a conceptual world within which truth-evaluable propositions about that world are available to you.

A second group of "expansive" or "liberal" naturalists acknowledge this need. John McDowell[81] and David Wiggins[82] have been at the forefront of these efforts. McDowell argues for the necessity of incorporating normative values that underpin epistemology, ethics, and aesthetics into a broader conception of "nature."[83]

John McDowell is one of the most significant philosophers of the late twentieth century, and his *Mind and World* has become a touchstone for debates concerning our relation to reality, the nature of experience, and the status of rationality.[84] The central target of his argument is a deeply entrenched picture of experience that has shaped modern thought since Descartes and is reflected in the ideas discussed above. As A. C. Genova succinctly puts it,

> With the advent of the new science . . . a certain picture of experience emerged. We now had the perceiving subject on one side and the disenchanted world on the other, and experience was now construed as both an externally caused effect on our sensibility and an occurrence having epistemological relevance for empirical belief. But how can this be? It appears that there is an incommensurability between the *normative* space of reasons . . . and the scientific domain of blind causal relations (nature). How can experience be a passive receptivity while also having rational relations to belief?[85]

to W. V. O. Quine.

81. See Lindgaard, *John McDowell.*

82. See Wiggins, "Neglected Position?"; Wiggins, "Cognitivism, Naturalism and Normativity."

83. Thornton, "Experience, Knowledge, and Openness."

84. McDowell, *Mind and World.*

85. Genova, review of *Reading McDowell*; emphasis added.

This picture generates an apparent difficulty: if experience is merely a causal impact of the world upon our sensibility, how can it also stand in rational relations to belief? The result is a supposed gap between "the normative space of reasons" and the "scientific domain of blind causal relations." For McDowell, however, this is not a genuine problem but a *pseudo-problem*, arising from a mistaken conception of both nature and experience. The tradition's attempts to resolve this tension, McDowell argues, fall into a series of unstable oscillations:

> The [traditional] philosophical responses to this . . . yielded two successive misguided dilemmas. The *first* is reflected in the oscillation between coherentism and the Myth of the Given. The former acknowledges an irreducible dichotomy between . . . rational relations . . . and the causal relations of nature . . . and opts for what McDowell characterizes as a frictionless system of normative relations among beliefs (only beliefs justify beliefs), thereby avoiding the problem but at the cost of no objective, external constraints—a disconnect between mind and world. [The Myth of the Given, on the other hand] construes experience as a foundational given that it has a *non*-conceptualized content [by this is meant a sort of unidentifiable or uncategorizable sensory input] in virtue of which it can somehow warrant empirical belief, thereby again avoiding the puzzle, but now, for McDowell, we have a *desideratum* that cannot be satisfied.[86]

So, on the one hand, coherentism preserves the normative domain by allowing only beliefs to justify beliefs, but at the cost of detaching thought from the world—what he calls a "frictionless" system. On the other hand, the Myth of the Given seeks to anchor belief in a nonconceptual experiential foundation, but thereby invokes something that, lacking conceptual form, cannot rationally justify anything at all.

This unresolved tension reappears in a second dilemma. "Bald naturalism" attempts to reduce or eliminate the normative in favor of purely scientific description, while "rampant Platonism" places the space of reasons outside nature altogether. Both, in different ways, accept the same underlying assumption: that nature is exhausted by law-governed causal relations and thus cannot itself be intrinsically rational or conceptually structured. McDowell's decisive move is to reject this assumption. As Genova summarizes,

> The [muddle] is due to a false picture shared by all of the above—the confused scientistic notion that nothing natural (including sensibility) can be shaped by *sui generis* conceptual capacities

86. Genova, review of *Reading McDowell.*

> because nature is exhausted by the *realm of law*. But this is a dogmatic superstition. . . . Experience is *already* conceptually informed; [the space of reasons] extends all the way out to the world—it is unbounded. Experience has conceptual content *as* a natural phenomenon. Thus, we recapture the correct picture via a modest re-enchantment of nature.[87]

On McDowell's view, then, our reception of the natural world by our senses and conceptuality are not opposed but integrated. In experience, the world is not merely causally impinging upon us but is directly present in a form that is already apt for rational engagement. Conceptual capacities are not imposed upon an alien natural substrate; they are operative within our natural responsiveness itself. The result is what McDowell calls a "modest re-enchantment of nature": a conception in which nature is not disjoint from reason but includes within it the possibility of intelligible, conceptually articulated disclosure.

The significance of this for the present argument is clear. If McDowell is right, then our normative responsiveness to the world—our capacity to recognize reasons, to judge, to be guided by evidence—is not something added to an otherwise nonnormative reality nor something inexplicably bridging two heterogeneous domains. It is instead a natural expression of our mode of being in a world that is already open to rational articulation. The relation between mind and world is therefore not external but internal to experience itself.

Within such a picture, our responsiveness to scientific evidence is no longer mysterious. To recognize something as evidence is already to stand in a normative relation to what is given, and this relation is made possible precisely because experience itself belongs within the space of reasons. The world, as it is disclosed in experience, is not normatively inert but already available to rational appraisal.

So, we can agree with Dawes that "evidence" is critical to theory but demonstrate that the concept of "evidence" which Dawes mobilizes as the distinguishing feature of MN is not immune from the need for an expanded picture of what is required for an acceptable naturalism.[88] It is this which McDowell provides.

"Evidence," as McDowell maintains, "represents" the world as being a certain way. He points out that there is a difference between a lawful causal relation between objects and the relation of "representing." Smoke is caused

87. Genova, review of *Reading McDowell*.

88. See Friedman, "Exorcising the Philosophical Tradition."

by fire, but fire does not "represent" smoke. As Mats Wahlberg describes McDowell's position,

> When we say that a belief is justified based on certain evidence, what we are talking about is a *normative* relation between the evidence and belief. By saying that the evidence justifies the belief we are not just saying that the evidence has *caused* us to have the belief (as the fire causes the smoke). We are saying something more, viz. that given the evidence one *ought* to have the belief.[89]

Every contributor to the discussion about science is likely to agree that that injunction is a distinctive feature of the sciences: the principle that laws, and theories, *ought* to be accepted and rejected solely with respect to the justice they do to the evidence. As I argued earlier, the application and understanding of *this* treasured idea require reference to concepts that are not themselves subject to potential scientific verification because they participate in understanding *what* such verification consists of.

What this makes possible, and what McDowell himself only gestures towards, is a further step that is central to the present work. If the world is disclosed in experience as already within the space of normative reason, then what is disclosed is not merely fact but *structured* fact—articulated, ordered, and available to judgment under norms such as coherence, unity, and adequacy. These are not external impositions but belong to the very intelligibility of what is given. But such norms are precisely those that, in their most developed form, constitute aesthetic appraisal. To recognize a structure as elegant, unified, or harmonious is not to add a subjective gloss, but to register the way in which it satisfies the very conditions under which it can be intelligibly grasped at all.

In this sense, the conceptuality that McDowell locates within experience already opens onto an aesthetic dimension: the world is disclosed not only as thinkable but as exhibiting forms that admit of better and worse realizations. The "modest re-enchantment of nature" thus deepens. Nature is not only conceptually available but normatively and aesthetically articulated, such that its comprehension and its beauty are internally related aspects of the same structured disclosure. It is the world offering to us the means not only whereby our judgments can truthfully or otherwise measure its value, but providing for us, in the nature of our cognitive capacity to be rationally involved with it, the means to do so—what we ought to do to acquire knowledge of its causal operations.

89. Wahlberg, *Reshaping Natural Theology*, 31.

Aesthetics and Normativity in Science

"Ought" involves a recommendation that steps away from factual description and *prescribes* the application of certain values *for* something to *count* as accurately representing some feature of the world. Hilary Putnam, the Harvard logician and philosopher of science, argued that "epistemic values, such as 'coherence,' 'plausibility,' 'reasonableness,' 'simplicity,' 'elegance,' . . . are presupposed in the activity of selecting scientific theories."[90]

Among these values are aesthetic judgments. Aesthetic judgments, as are the others, are normative; their deployment requires reference to some prior standard of what is to count as a "beautiful object" and what is not, and all the options, even the "hazy" and "vague" ones, in between.

Every single one of Putnam's presuppositions carries a demand for an evaluation understood either in respect of mastery of the application of a concept and/or simultaneously with some idea of what, in each case, is going to count as a positive or negative evaluation of the feature in question.

Curiously, the reader will recall, Dawes himself admits that there are "commonly accepted lists of explanatory virtues"; "virtues," of course, are not only distinctly normative but implicitly commendatory. Among those he lists, the reader will recall, are "scope, explanatory power and simplicity, or testability, ontological economy and informativeness." Not only are these criteria offered as a list of demands that ought to be followed in decision-making, but most of them involve an explicit appeal to principles that *could* imply, at least, that the structures of nature might be characterized by the qualities they invoke, "ontological economy," "simplicity," and so on.

My suspicion is that, like many who have absorbed the personal/nonpersonal distinction and its accompanying doctrines, such as the "fact–value" distinction and the subjective/objective epistemological distinction, it seems entirely intuitive to him that such things as "aesthetic properties," "truth," and other such characteristics of theory evaluation, "ontological simplicity," etc. are not properties of the *intrinsic* structures of "natural world" as the two-world metaphysics assumes.

As I have suggested, however, normative properties are in fact conditions on *anything* appearing as an "intelligible structure" available for scientific investigation. Insofar as this is the case, then *any* natural structure open to empirical investigation will necessarily display them. Aesthetic properties, for example, I will suggest, are a necessary to the appearance of anything as a structure. *What* aesthetic properties the natural world manifests is a matter

90. Putnam, *Fact/Value Dichotomy*, 141.

for empirical investigation, but *that* they display such properties *at all*, and *do* display them for empirical investigation, is a metaphysical necessity.[91]

"Imaging" the World: Theory as Structural Representation

The sorts of considerations raised by McDowell and others helps motivate a shift from the idea that normative features such as aesthetic properties could only be "subjective" in nature, having no bearing on the proper relation between a theory and reality nor on the nature of that reality itself.

Some contemporary authors have begun to recognize that how theories are developed in science bear many important relationships to how artistic creations emerge from the plenitude of human creative faculties, including structural and mathematical imagination.[92]

A recent work exploring the richness of this relationship is Tom McLeish's book *The Poetry and Music of Science: Comparing Creativity in Science and Art*.[93] One subsection in chapter 3 is titled "Mathematical-Theory Painting." This heading alone is evocative in the picture it creates of mathematical attainment involving reference to the normative features of artistic creation in mathematical proof; it invokes the idea that formulating a proof in mathematics is sometimes as much a business for mathematicians as finding a satisfactory structure to proof via creative processes is the straightforward business of logico-mathematical analysis.

The place that "subjectivity," "beauty," "imagination," and so on have in the sciences has typically been restricted to the development of initial ideas and hypotheses.[94] It is encapsulated in logical positivist philosopher Hans Reichenbach's famous division between the *context of discovery* and the *context of justification*.[95]

91. Ellis, *God, Value, and Nature*, offers a sustained defense of the idea that reductive naturalism should be rejected in favor of the more "expansive" naturalism associated with McDowell and David Wiggins. She argues that this "expansive" naturalism moves toward the idea of God while still allowing us to remain naturalists with respect to empirical science, since there is no logical incompatibility between the two. This is also broadly my own position. Ellis, however, provides a comprehensive book-length philosophical examination of naturalism and "expansive" naturalism to which I cannot do full justice here.

92. Wechsler, *On Aesthetics in Science*, provides case examples of the use of aesthetic criteria in science, ranging from imagery to scientific models.

93. McLeish, *Poetry and Music of Science*.

94. Kekule's structure for benzene, for example, is reputed to have come to him in a dream.

95. Reichenbach, *Experience and Prediction*, 38.

In this classic picture, the two contexts of scientific activity are differentiated by the supposed *arrationality* of the former and the *rationality* of the latter. Intuitions, imagination, "leaps of faith," a preference for the beautiful, and all manner of cognitive and evaluative phenomena may be factors in the *generation* (the context of discovery) of hypotheses and experimental tests. Their final appraisal and assessment in the context of justification, the province of rigorous empirical method and experiment, bears no important relationship to this ultimate satisfying of epistemological verity by method. Whether the boundary between these two distinctions is a clear dichotomy rather than a useful methodological distinction is cast into doubt by works by authors like McLeish.

The shift at least of the importance of factors such as those involved in creation must be also associated with the idea of structure in its broadest form, including the structures of mathematics. This is a question I take up in the next chapter. Although I cannot fully pursue the idea here, it is probably not a coincidence that the growth of interest in creativity in arts and science is also associated with a shift in the middle and late twentieth century away from "positivism" in the philosophy of science,[96] to which Reichenbach's distinction contributed.

"Positivism" essentially asserted that if anything were to count as "scientific" knowledge, it should be possible to display it in the form of a deductive argument with empirically verified premises leading to a deduced conclusion expressed in formal logic. "Theories" are thus understood as sets of logically related propositions, enabling their assessment for truth or falsity. It is in essence to distill scientific theory into a deductive form to provide analytic clarity to the epistemological claims of the natural sciences.

The shift away from this picture of scientific knowledge involved, among many developments, the *model-theoretic* approach to understanding science, versions of which advanced the idea that theories are "representations,"[97] or attempts at modeling parts of nature.

While the positivist hope embodied the ideal of a set of abstractly defined logical relations on the basis of observational premises, the general concept of "representation" engaged a fuller range of properties of theories congruent with human intellect, imagination, and perception in the cognitive engagement with a "natural world."

Ivanova and French describe the primary change from the syntactic approach of positivism to a richer representational account:

96. See Ivanova and French, *Aesthetics of Science*, 1–20.

97. The locus classicus for this change is Suppes, "Meaning and Uses of Models."

> How theories represent the world became the central question, with many commentators drawing analogies with the representational nature of artworks and scientific theories explicitly compared to such artworks.[98]

Naturally, these developments have seen an increasing focus on aesthetic properties and their relation to theories, the properties of the world that such representations may reflect, and the relation of truth to aesthetic value.[99] This engagement with aesthetics is in the context of a wider acknowledgment of the place of value and normativity in the constructions of scientific activity.

If this is right, then the natural relations between mathematics, structure, and aesthetics enter as metaphysical participants in describing what constitutes the propensity of the natural to be "the natural," visible to empirical investigation, and rationally intelligible. The object of scientific investigation is a structure normatively bound by the aesthetics it expresses. What makes it conceptually and perceptually available to human subjects are the theoretical representational structures which bear mimetic normative relations to the physical events and relations they are attempting to discern.

This is nicely illustrated by a (common) visual picture of one of the great achievements of physics: the Scottish scientist James Maxwell's equations describing the unity of electricity and magnetism in wave form.

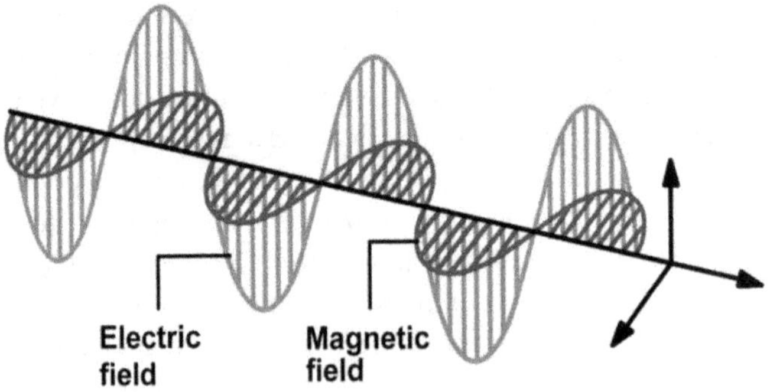

A representational form of James Clerk Maxwell's electromagnetic wave

The representation involves both visual and abstract components brought together by the imagination to create an artifactual object which has a

98. Ivanova and French, *Aesthetics of Science*, 109.

99. See Bueno et al., *Thinking About Science*. This source contains a series of articles on, among other things, shared properties of representation in art and science and how these are to be understood.

scientific purpose. While it might be thought that the visual representation is only analogous in form, it clearly shares with Maxwell's theory and equations structural features that transcend both of these and are those in virtue of which the analogy can be scientifically instructive.

So, for example, it is clear that certain statistical shapes repeat in a pleasing fashion, that they represent two different kinds of movement satisfyingly orthogonal to each other but united in the direction of flow. What is less obvious in the visual display is the extent to which shapes and structures considered in their purely physical dimensions might also exercise a restraint in what analogically or representationally can be physically displayed. Not any physical shape will do, in other words. One cannot represent "chaos," for example, by the Maxwellian representation above.

As this example demonstrates, by a happy isomorphism, human cognitive abilities like perceptual and abstract imagination, otherwise rejected by the propositional strictures of positivism, become detectors and analysts of aesthetic and other structural elements.

In the following chapter I suggest that the structures of nature at the most basic levels of mathematical physics do indeed express aesthetic qualities. In fact, I argue that it is a metaphysical principle of anything that exists that it has structure, and that "structure" is not metaphysically decomposable from its expression of some aesthetic property.

Both of these requirements I suggest engage normative evaluation from physicists that can contribute to theory assessment. I present the case and evidence for the empirical manifestation of "beauty" in these depths of the physical. This is a first step towards reunifying teleology, normativity, and the personal with the impersonal.

I end this chapter with a quote from Brooke and Cantor, historians of religion and science. In a chapter of their book *Reconstructing Nature: The Engagement of Science and Religion*, they aver that

> an uncovering and appreciation of beauty in the scientist's construction of nature have been prominent goals of scientific enquiry, with edifying . . . effects. . . . Edifying because it helped to bring out harmonies and beauties *in* nature that might have otherwise been missed.[100]

The world of beautiful sub-particles and forces found by physics to make up the basic physical ontology of the world are empirical affirmations of this assertion. I justify and expand this conclusion in the next chapter.

100. Brooke and Cantor, *Reconstructing Nature*, 109.

Chapter 2

Structures in Mathematical Physics

A World of Phenomenal and "Measurable" Beauty

SUMMARY

I OFFER AN ACCOUNT in this chapter of how the basic structures of the physical world investigated by particle physics can be properly understood as uniting subjectivity, normative properties like aesthetics, and mathematical measurement, without threat to empirical science.

OVERVIEW OF THE CHAPTER

The general idea of "structure" I aim at defending in this chapter is that it is a prerequisite of anything coming into physical (or abstract) existence. The idea has evolved into a sophisticated and full-blown program in the philosophy of science known as "structural realism." This is a view postulated early on in the twentieth century by, among others, the British physicist Sir Arthur Eddington, who held that mathematical structures constituted the base elements of the world.[1]

The relevance of this to my central theme is that if such structures are fundamental, then the question immediately arises as to how they are to be understood: whether purely in terms of objective, measurable features, or whether they may also legitimately be said to exhibit expressive and normative properties.

1. Eddington, *Nature of the Physical World*.

The properties of mathematical structures, which are often held to be acausal, abstract, and outside space-time, coupled with the views of physicists who take them to be fundamental, present a problem for philosophical naturalists. For if the most basic constituents of reality are of this structural and mathematical kind, it becomes unclear how they are to be accommodated within a conception of the "physical" that excludes expressive and normative features. The "physicalist" naturalist would like reality at base to consist of objective, measurable, physical, and preferably observable properties in regular causal relations, and possessed of no interesting expressive or normative features. At the very least, the fundamentality of mathematics in particle physics places pressure on what is meant by "physical."

I offer an account of the relations between structure, mathematical physics, and aesthetics that allows for a conception of fundamental "physical" reality capable of incorporating expressive and hence normative features. While mathematics describes the objective quantitative dimensions of structures, the expressive properties that they manifest aesthetically open them up to forms of judgment that are irreducible to measurement alone.

Despite the involvement of subjective experience, and the variance it can introduce, the truth-aptness of such judgments is not relative to the person who makes them. The reason for this is that although an essential feature of any aesthetic experience is phenomenal, aesthetic properties are not optional additions to structures. Rather, anything that has a structure must, in virtue of that structure, admit of aesthetic characterization. I claim that there are different but epistemologically legitimate modes of access to the aesthetics of structures, both "personal" and "impersonal."

In other words, while aesthetic judgments of these arcane structures of basic matter may be informed by their mathematical construction, they are also subject to forms of personal discernment. This does not disqualify them from being real features of the physical world. They can be experienced and used for theoretical judgment by a mathematical physicist fully acquainted with their discipline and possessing a high degree of cognitive and imaginative sensitivity to the aesthetic features manifested in the structures of mathematics and physics.

A simple example of this may be found in geometric practice: a particular ratio of a part of a rectangle to the whole is often the ground of aesthetic responses of beauty. Such cases provide a model for a working epistemology in which aesthetic apprehension is neither arbitrary nor reducible to purely formal description.

This, I argue, involves an appreciation of how the various components of structure—such as symmetry, unity, and simplicity—are arranged so as to produce an aesthetically intelligible theoretical order. "Order" in this sense is not merely descriptive but plays an important role in theory evaluation. I offer two examples of this in practice: the first demonstrates the central function that symmetry plays in physical theory; the second discusses the Nobel Prize–winning physicist P. A. M. Dirac, who asserted that it was more important to have beauty in one's equations than for them to agree, initially, with experimental findings. The significance of Dirac's claim lies in his explicit use of the term "beauty" to refer to the phenomenology of its experience as a guide to the likely truth of a theory.

Finally, I provide a range of examples of physicists and mathematicians who have made it explicit that they find beauty in the basic structures of nature, and I give a brief indication of how Einstein's theories of relativity have been widely regarded as exemplifying such beauty.

PART 1

A World of Structures

The 1998 Darwin Lectures at the eponymous Cambridge University College gave rise to a collection of articles published by Cambridge University Press entitled *Structure in Science and Art*. In her opening discussion of the title, Wendy Pullan introduces the topic as follows:

> Bridges and buildings, DNA and the periodic table, flora and fauna, machines and circuits, human beings and the societies of which they are part, even thoughts and ideas—all of these can be understood as particular structures that are part of the world and our experience of it.[2]

What is immediately striking about this characterization is its breadth. The concept of structure is not confined to any one domain of inquiry nor to any one kind of object. It ranges across the physical, biological, social, and conceptual, uniting under a single term phenomena that differ widely in scale, composition, and mode of existence. What these disparate cases share is not a common material basis but the fact that each involves elements related to one another in determinate ways. To describe something as a structure is, at the most general level, to say that it is not a mere aggregate but an organized whole.

One deep thread in the etymology of the word "structure," and hence "construction," is the Latin *structura*, a "putting together." The concept is as old as Western philosophy and science, and its persistence reflects

2. Pullan, "Structuring Structure," 1.

something fundamental about the way in which the world is encountered. Pullan notes this and expands the idea: "Certain words spring to mind: organisation, arrangement, connection, orientation, framework, and order."[3]

These associated terms are not accidental. They point toward a common underlying idea: that for anything to count as a determinate object or system, its elements must stand in some ordered relation to one another. This suggests that the concept of structure is not merely classificatory but may be elevated to the level of an a priori principle. One may argue that structure is fundamental to any conception of a world that can be experienced, that is contingent, and that is open to investigation.

If this is so, then structure is not one feature among others within the world, but a condition of there being a world at all. To encounter something as an object, as having identity, persistence, or properties, is already to encounter it as structured in some way. A world devoid of structure would not be a disordered world; it would not be a world in any intelligible sense. Without some principle of organization, there would be no differentiation, no identifiable entities, and no relations capable of description or investigation.

Of course, the concept of structure is associated with a number of auxiliary metaphysical considerations. There may be certain ways in which the relationships between the parts and wholes of a structure are necessarily conceivable. No structure, for example, could exist as a whole prior to the existence of its parts, even if the whole exhibits properties that are not reducible to those parts taken in isolation. Similarly, functions, shapes, dimensions, and causal processes must conform to principles, often mathematical in form, if they are to constitute logically possible structures. These modal constraints are not derived from observation but are presupposed in the very idea of a configuration that can exist.

At the same time, care must be taken not to assume that concepts such as "order" and structure imply fixity, rigidity, or completeness. As Pullan emphasizes, "Structure can be open and dynamic, a matter of structuring as much as structure."[4]

This qualification is essential. The world is one of constant change and causal activity, and the task of the sciences is to provide theoretical representations of the functioning of the structures in virtue of which this activity occurs. Structure, in this sense, is not opposed to process but is the condition under which processes take determinate and intelligible form.

3. Pullan, "Structuring Structure," 1.

4. Pullan, "Structuring Structure," 1.

Consider, for example, the case of water. It is composed of hydrogen and oxygen in polymeric forms such as $(H_2O)_2$ and $(H_2O)_3$. These elements are in a continual state of structural formation and dissipation. Bonds are constantly breaking and reforming, yet this flux exhibits a determinate organization that gives rise to the familiar macroscopic properties of water. What is observed here is not the absence of structure but its dynamic realization.

The point is made here in broad terms, but it is sufficient to indicate the rough metaphysical outline of one aspect of the natural world. David Pettifor and Alan Cottrell in their chapter "Models of Structure," in the aforementioned volume,[5] offer the view that all of the sciences may be understood in terms of three main types of structure: the cosmological, or gigantic; the "miniscule," dealing with basic particles; and the highly complex, encompassing disciplines such as biology, geology, condensed matter physics, chemistry, and the social sciences, including economics and psychology.

The precise relations between these domains are, of course, complex and contested. For present purposes, however, the Pettifor and Cottrell distinction serves as a useful initial framework. Whether or not it requires refinement, it supports the more general claim that structure is essential to the manifestation of anything that might be said to exist physically and to be the subject matter of scientific inquiry.

The conclusion to be drawn at this stage is therefore limited but fundamental. Any world that can be experienced, any object that can be identified, and any process that can be investigated must exhibit structure. This is not a contingent feature of our world that might have been otherwise, but a condition of there being anything that could count as a world at all.

The question that now arises is whether anything further follows from this. In particular, if structure is a necessary condition of intelligibility, does it carry with it additional features that are equally inescapable?

The Necessary Aesthetic Character of Structures

If the foregoing philosophical proposals are plausible, one necessary consequence follows from them. Carol S. Gould describes it thus:

> If one accepts the existence of any world, then . . . the existence of at least one aesthetic property must be granted: the world

5. Cottrell and Pettifor, "Models of Structure."

> must exemplify some arrangement of its constituents and any arrangement of features must satisfy some aesthetic predicate.[6]

A natural first response to such a suggestion is to insist that whether or not this is the case is a matter for empirical exploration rather than metaphysical declaration. Since the claim concerns the structures that determine the nature of the things around us, the sciences of nature might be thought to be the appropriate arbiters of whether or not they express aesthetic properties such as beauty. On this view, aesthetic features, like any other characteristics attributed to the physical world, are contingent. They need not exist, contrary to Gould's assertion, and whether they do or not is to be settled a posteriori.

A proposal of this sort, however, confuses two issues of considerable importance. Gould's point is not that metaphysical priority dictates what aesthetic features any structure must express. It is rather that anything which has a structure, and the world consists of structured things, must, in virtue of that structure, admit of some aesthetic characterization. The necessity lies not in the specific form of the aesthetic property but in the fact that some such property must be attributable.

What these properties are is a matter determined by the nature of the structure in question. That determination is indeed a posteriori, reserved for empirical observation or experience, whether the object concerned is abstract or "substantial." Thus, whether the ultimate structures and processes of biology or particle physics turn out to be "ugly," "pretty," or aesthetically indifferent is a question for the relevant scientific disciplines. The same applies in mathematics: whether a proof is elegant or cumbersome is often a matter for those with the requisite expertise to judge.

Whether any abstract or perceptible structure has some aesthetic character at all, however, is not a matter for any discipline. It is a prerequisite of there being any organized matter, abstract object, or process for those disciplines to investigate in the first place. The distinction, therefore, is between the *necessity of aesthetic character as such and the contingency of its particular form.*

The philosopher Eddy Zemach describes the empirical situation in the following way: "Even if a description of reality as it is did not include any predicate of our present science, it cannot fail to include aesthetic

6. Gould, "Reality of Aesthetic Properties," 350.

predicates."[7] As he goes on to say, "If it has some nature, it has a character and structure."[8]

In other words, should every current scientific theory we have fail, whatever it was that replaced them would still of necessity have aesthetic features.[9] There is something fundamental, then, to any aspect of a world—whatever form its existence takes—that is inescapably aesthetic. Any structure will have parts arranged in certain ways, and those arrangements will contribute to the evaluative judgments they elicit. If this is correct, it is to be expected that those structures which feature in scientific theories will be no exception.

This expectation finds support in scientific practice itself. In the previous chapter I noted that the ontological naturalist Gregory Dawes admits predicates of a distinctly aesthetic character into the theoretical virtues by which competing theories are assessed. When such predicates are examined, their connection to structure becomes clear. "Unity" suggests the bringing together of disparate elements into a single framework; "harmony" refers both to a mathematical description of musical relations and, more generally, to the coordinated functioning of components; "symmetry" denotes the invariance of relations under transformation—a rotated circle, for example, remains the same.

These are not incidental features. They describe ways in which parts are related within a whole, and as such they belong directly to the domain of structure. Their aesthetic character is not an external projection but arises from the organization of the relations themselves.

The conclusion to be drawn is that aesthetic features are not confined to particular domains, such as art or subjective experience, but will show up in all aspects of the world, including those investigated by the natural sciences. If so, any discussion of beauty in natural phenomena must engage with the relevant scientific disciplines in order to determine what aesthetic features are present and what role they play within the theoretical structures those disciplines employ.

7. Zemach, "Real Beauty," 262. He also offers a full-length book in philosophical defense of the idea that aesthetic properties are real properties of things against opposing views emerging from subjectivism, noncognitivism, and relativism. All these positions are well established across a range of philosophical questions including ethics, the perception of reality, and truth, and in one way or another deny the existence of, or availability of, objective properties or objective truth in different or all domains of philosophical interest; see Zemach, *Real Beauty*.

8. Zemach, "Real Beauty," 262.

9. Budd, review of *Real Beauty*, offers an outline of Zemach's work together with a critique.

Aesthetics, Beauty, and Physical Foundations

There are philosophical and theological reasons for particularly hoping that beauty will be one of the characteristic aesthetic features that underlie the structures of the natural world. The discipline associated with the forces and elements that constitute the entirety of the world as it presents itself to us is physics. The discovery of beauty within this discipline, and hence within the universe itself, would provide empirical support for the claim that the universe has a particular aesthetic character at its base, as feature of the structures under investigation. It might even support the assertion that beauty is necessary for the base foundations of a physical universe, an empirical, therefore gainsayable, proposition perhaps, but the closest one might come to a non–a priori truth about its aesthetic nature.

As I have argued, all structures are of necessity aesthetic in their composition. This, as I pointed out earlier, is sufficient on its own to provoke the idea or even the direct perception of intentional creation, but should it be the case that physics evinces its beauty as a matter of empirical discovery, then such a finding would provide a not only a further persuasive basis for the claim that the heart of the world manifests something other than brute causation and empty physicality, but that it was meant to be beautiful. The consequences of this for our perception of a creator and his intentions for the world are not part of my discussion but surely would manifest goodness rather than malign intent. Why beauty, after all, rather than aesthetically dull or indifferent structure?

The range of structures dealt with by physics is vast, and it is important to convey this breadth without becoming lost in technical detail. Physics is commonly defined as the science that investigates matter, energy, and their interactions, but this definition scarcely captures its scope. What distinguishes physics is that it seeks not merely to describe particular phenomena but to uncover the most general principles that govern all phenomena whatsoever. It operates, as it were, at the deepest level of structural description available to us.

At one end of its range, physics concerns itself with the most immediate and familiar features of the world: motion, forces, heat, light, and the behavior of physical systems that we can observe directly. At another, it reaches into domains that are far removed from ordinary perception: the subatomic world of fundamental particles and fields, and the cosmological scale of the universe as a whole. Between these extremes lie the countless structures that give rise to the material forms of everyday experience—solids, liquids, gases, biological systems, and the complex organizations studied by the special sciences.

What unifies these diverse domains is not their subject matter taken in isolation but the fact that they are approached through a common aim: to identify the underlying structures that render them intelligible. Physics does not simply catalogue events; it seeks to express the relations governing those events in precise, often mathematical, form. In doing so, it provides the framework within which other sciences situate their own more specialized accounts. It is for this reason that physics is often regarded as a foundational discipline. It does not exhaust the description of reality, but it sets the terms in which many other descriptions must be framed.

This centrality explains why physics forms the primary arena for the present discussion. If the claim is that structure is fundamental, and that aesthetic character is necessarily implicated in structure, then the discipline that most directly investigates the fundamental structures of the world is the natural place in which to test that claim. Moreover, if aesthetic properties are to be found at this level, they cannot easily be dismissed as superficial or merely human projections. They would belong to the very fabric of the structures that physics describes.

My purpose in sketching this range is not to provide a technical survey but to indicate the breadth of the domain within which the relations between mathematics, aesthetics, structure, and metaphysical necessity must be considered. Within each area of physics, the precise form of these relations will be locally determined. But that they can have any determinate form at all, as I have argued, depends upon the more general principles already established. At the same time, the subject matter of physics provides an important arena in which the claims of reductive naturalism may be tested, and in which questions concerning creation, God, and the nature of reality come into view.

My brief excursion into the domain of physics should therefore be sufficient to alert the reader to the substantive breadth of these proposals. Let me now turn to some of the reasons that underpin them. First, it is, perhaps paradoxically, physics itself that challenges the claim of reductive naturalism, such as that advanced by Dawes and others, that all the items and properties of the world either reduce to, are supervenient upon, or are realized by purely physical entities, where this excludes normative and expressive characteristics.

The problem usually taken to confront such naturalisms concerns consciousness, intentionality, and teleology: how to accommodate these within a physical world that, by definition, excludes them except as derivative or dependent features. This is often framed in terms of the relation between mind and body. There is, however, another difficulty, arising from within physics itself, that generates a more fundamental challenge. It is this: basic or fundamental physics is deeply mathematical.

Despite the almost intuitive appearance this assertion now has, it was not always thus. Aristotelian natural philosophy conceived mathematics to be the business of its own discipline and separate from the activity of observing the natural world and drawing principles of general explanation from those observations. The development of the application of the theoretical application of mathematics to the natural world and physics came later.[10]

The Merton Calculators, for example, a group of fellows of the eponymous Oxford College, were part of a general movement including the university at Paris from the late thirteenth to fourteenth centuries who began to develop ways of "using the idea of a mathematically universal law to understand the physical world."[11]

Thomas Bradwardine, for example, a fellow, then professor of divinity, chancellor of Oxford, and archbishop of Canterbury at the end of his career, expressed the significance of this development in striking terms:

> Mathematics is the revealer of every genuine truth. . . . It knows every hidden secret and bears the key to every subtlety of letters. Whoever then has the effrontery to pursue physics while neglecting mathematics should know from the start that he will never make his entry through the portals of wisdom.[12]

This conviction finds its classical expression later in the work of Galileo Galilei (1564–1642), who, in developing methods for measuring quantities such as velocity and acceleration, famously wrote,

> Philosophy is written in this grand book, the universe, which stands continually open to our gaze. But the book cannot be understood unless one first learns to comprehend the language and letters in which it is composed. It is written in the language of mathematics, and its characters are triangles, circles, and other geometric figures without which it is humanly impossible to understand a single word of it.[13]

10. Hobart, *Great Rift*, takes the view that the development of mathematics and symbolic representation caused a "rift" between science and religion. He thinks that this development represents two different ways of seeing the world.

11. Wagner and Briggs, *Penultimate Curiosity*, 172.

12. Molland, "Geometrical Background," 110.

13. Galilei, *Assayer*, 237–38.

Some contemporary physicists and mathematicians have gone further, proposing that the basic structures of the universe are themselves mathematical,[14] a pronouncement which throws the physicalist enterprise of casting the ultimate ontology of the world into a model of physicalist entities into serious difficulty. *Whatever* the correct assignation of mathematics is to the fundamental nature of the physical world, the necessity of the description of those events in mathematical terms provokes a crisis for the philosopher who wishes to garland events conceived as "physical" with the accolade of ontological fundamentality.

The problem may be stated as follows. Mathematical entities are typically regarded as acausal, outside space-time, and nonmental. Yet physics, which is often taken to provide the strongest empirical support for a purely physical ontology, depends essentially upon such entities for its formulation. How, then, can physics be invoked as evidence for the claim that the universe admits of no ontological space for nonphysical realities when its own theoretical structure appears to rely upon them? At best, this places the physicalist position under considerable strain; at worst, it suggests a tension within the position itself. Susan Schneider, in her discussion "Does the Mathematical Nature of Physics Undermine Physicalism?,"[15] refers to this as the "Problem of the Base." It presents prima facie difficulties for any ontological naturalism that insists upon a reductive physicalist foundation.

A second reason for taking physics as the primary arena for the discussion of beauty lies in its account of the origins of the universe. Cosmology and particle physics converge upon the view that the universe, including space-time itself, emerged approximately 13.8 billion years ago. Within an extraordinarily brief interval following this emergence, the fundamental forces of nature came into operation, and the basic constituents of matter began to form. From these initial conditions, through processes of increasing complexity, there arose atoms, stars, chemical structures, biological systems, and eventually conscious life.

If the argument advanced in earlier sections is correct, then these earliest structures, those that mark the coming into being of the physical

14. See, for example, Tegmark, *Our Mathematical Universe*. Kragh, "Mathematics and Physics," summarizes the history of what he calls the development of a "strongly mathematized physics," including figures such as the mathematicians David Hilbert and Hermann Minkowski, together with physicists including Albert Einstein, Arthur Eddington, and Max Tegmark. Farmelo, *Universe Speaks in Numbers*, offers a sustained defense of the view that contemporary mathematical physics, in continuity with the tradition of Newton, remains central to our understanding of deep reality, against commentators who argue that physics has "lost its way" in mathematical fantasies bearing little contact with physical reality.

15. Schneider, "Mathematical Nature of Physics."

world, must already have possessed some aesthetic character. The question, therefore, is not whether aesthetic properties are present but what form they take. If, as many physicists have suggested, the underlying structures of the universe exhibit forms of simplicity, symmetry, or elegance, then the beauty of its aesthetic dimension is not a late or superficial addition to reality but is present at its very origin.

A third reason concerns the proximity of physics to metaphysical questions. It is within this domain that some of the most fundamental issues concerning the nature of physical reality are brought into sharp focus. Developments within physics have repeatedly challenged our ordinary conceptions of identity, substance, and causation. The behavior of light, for example, exhibiting both particle-like and wave-like characteristics under different conditions, strains conventional categories of classification. Quantum theory, in its account of subatomic phenomena, raises further difficulties: What is the nature of the entities described by its formalism? Are they particles, waves, fields, or something else entirely?

Such questions are not merely technical. They bear directly upon the kinds of ontology that can be sustained. It is in this context that some theorists have been led to suggest that the ultimate nature of the universe is best understood in mathematical terms. Whether or not this conclusion is correct, it reinforces the centrality of structure, and of mathematical structure in particular, to any adequate account of reality.

The overall question that these observations provoke is therefore this: How are we to conceive of the general relations between mathematics, the physical universe, structure, and aesthetics? The first step is to consider how physics understands the place of structure in its account of the natural world, and how mathematics functions within that understanding.

This is a topic of vast importance and ongoing debate. It is fundamental to our grasp of the nature of the world, yet it remains unresolved. Eugene Wigner captured the difficulty in the title of his well-known paper, "The Unreasonable Effectiveness of Mathematics in the Natural Sciences."[16] The phrase points to a widely acknowledged fact: that mathematics is not merely a convenient tool but appears to stand in a deep and perhaps unexpected relation to the structure of physical reality.

At a minimum, it may be agreed that mathematics provides our most powerful means of articulating the structures of fundamental phenomena. I take from Wigner's heading the justification for assuming that at some bare minimum all are agreed that, firstly, mathematics does encompass some basic relation to the structural nature of fundamental things—at least

16. Wigner, "Unreasonable Effectiveness."

it is our best option. The physicist Murray Gell-Mann, who won the Nobel Prize for postulating the existence of "quarks," the ultimate buildings blocks of the physical world, put it this way: "How can it be that writing down a few simple and elegant formulae . . . can predict universal regularities of Nature?"[17]

Whatever the ultimate ontology to which mathematics relates, it is difficult to deny that it engages, in some sense, with the structures that constitute the physical world.

Aesthetic Emergence, the Origins of the Universe, and Physics

Let me conclude the section by summarizing the argument briefly, and in logical format, and rehearsing what follows from it. It begins with the following question: Are there any metaphysical claims that can be made about the coming into existence of the structures of the universe that are consistent with empirical discovery, yet not reducible to it? In particular, can the relation between structure and aesthetic character, as articulated by Gould and Zemach, be extended to the earliest phases of the universe described by contemporary physics?

The following line of reasoning suggests that this may be so:

1. Everything that comes into existence, including space-time, has a structure.
2. Every such structure is either mathematical in itself or admits of mathematical description.
3. Everything that has a structure necessarily admits of aesthetic characterization.
4. Whatever admits of aesthetic characterization is open to normative evaluation.

From these premises, the following conclusion may be drawn:

Every structure that comes into existence must have a mathematical character (or description) and at least one normative property—namely, an aesthetic one.

17. Gell-Mann, "Banquet Speech." Not everyone is agreed that the universe *is* structural-mathematical in some sense of that phrase. Stephen Wolfram, who created the symbolic language *Mathematica*, argues that the universe only *looks* mathematical. He suggests, for example, that not everything can be given a mathematical description. Complex phenomena like biology and weather turbulence, he suggests, have no descriptive mathematics. See Wolfram, *New Kind of Science*.

The significance of this conclusion is considerable. If correct, it implies that at the point at which the physical world comes into being, it does so under conditions that already include an expressive, evaluative dimension. The emergence of space-time and matter is not the emergence of a purely neutral or value-free domain, but of a reality structured in such a way as to admit aesthetic character and judgment.

The subject matter of particle physics, which concerns the earliest physical constituents of the universe, must therefore include at least one normative property—an aesthetic one—among its theoretical commitments. The empirical question that remains is what form this aesthetic character takes: whether it is realized in beauty, ugliness, or some more complex configuration.

If aesthetic properties are not only necessary features of structured existence but are present at the very origin of the physical world, then a further question arises: Why should this be so? Even if it is granted that the ultimate ontology of the universe consists in mathematical structures rather than material "stuff," those structures will still possess aesthetic character. If they are found to exhibit beauty, then this fact calls for further explanation, one which cannot be exhausted by the sciences, or so I will contest.

It is this line of thought that will be developed in what follows. The next step is to establish more fully the plausibility of the claim that structures are ontologically fundamental, and to articulate an account that unites structure, mathematics, aesthetics, and the epistemological accessibility of structures to us, both through their mathematical articulation and their aesthetic expressiveness.

Structural Realism and Mathematics in Physics

The British physicist and Quaker A. S. Eddington (1882–1994), who provided one of the early confirmations of Einstein's general relativity theory by his demonstration that light does indeed bend in the presence of a large mass, held the view that

> the investigation of the external world is a quest for structure rather than substance. A structure can best be represented as a complex of relations and relata.[18]

18. Eddington, *Mathematical Theory of Relativity*, 41.

In his magisterial book[19] on the relation between Eddington's science and his Quaker convictions, Matthew Stanley comments on the work in which this quote appears by stressing the basic point,

> He [Eddington] had made his argument for the structural basis of physics before, but this time it was inextricably woven into his physical reasoning and mathematical deductions. It was not merely an interpretation layered on top of physics. It was wholly integral to the science. *Structural considerations made the laws of physics inevitable.*[20]

In contemporary mathematical physics it is commonplace to talk about a "structure" posited by some theory; Einstein's special relativity theory is a theory about space-time structure. The structure of space-time in special relativity is said to be "Minkowskian," after the mathematician who derived a set of equations to describe it.

This position has developed into a view in the philosophy of science called *structural realism*,[21] one that argues for structure as the fundamental ontology of the world. J. Worrall, who is conventionally taken to be the locus classicus for contemporary structural realism, developed the view in response to the conflict between *realism* and *anti-realism*.[22] To understand the basic claim of structural realism, we have to understand the problem it tries to meet.

Anti-realists about physics hold that the discontinuities of theory change in physics do not provide evidence for a "scientific image" of an

19. Stanley, *Practical Mystic*. Interestingly, Stanley considers that Eddington's Quakerism and its commitment to the idea of "seeking" (Quakers originally identified themselves as "Seekers") provided him with a theoretical value that stressed an openness to theoretical and methodological exploration, thus preventing an attachment to one theory or method as sacrosanct.

20. Stanley, *Practical Mystic*, 182.

21. As one might expect, variants of structural realism have proliferated. The two principal divisions are between those who argue that "relations" are the primary ontological features of structures and those who maintain that "individuals" are the primary ontological elements within structural realism. A further distinction is made between "ontic" structural realism, which grants ontological primacy to structures themselves, and "epistemic" structural realism, the view that all we can ever know are structures. The literature on these issues and their many variants is now highly sophisticated. A good overview is provided by Ladyman, "Structural Realism."

Psillos, *Knowing the Structure of Nature*, offers a defense of scientific realism and epistemological optimism concerning the prospects of science for understanding nature.

22. Worrall's "Structural Realism" is the locus classicus for this view. In later years it has been developed particularly by Steven French and James Ladyman. See, for example, French, *Structure of the World*, and Ladyman, "What Is Structural Realism?"

underlying "physical" or "natural" reality uncovered by physics. "The quantum image, for example, of the world is totally different from the classical image of the world."[23]

We cannot rationally guarantee of any theory in physics that it has finally uncovered an independent reality; this is known as the problem of *meta-induction*—most theories have been superseded, so relying on inductive evidence to bolster the case that physics slowly uncovers something called "reality" is suspect.[24]

On the other hand, the provision of scientific explanations seems to provide useful, productive, and empirically supported means of manipulation of the external world about us, which does suggest the uncovering of a reality independent of us. This is commonly referred to as the "no miracles argument," which is intended to imply that without the notion that science deals in an external reality its success would be "miraculous."

Many of the great achievements of physics seem to have involved progress in uniting different phenomena in mathematico-theoretical advances: the bringing together of the movements of the objects of the heavens with the movements of the objects of the earth and the earth itself[25] (Isaac Newton, 1642–1726); light, electricity, and magnetism into one set of equations (James Clerk Maxwell, 1831–1879); and, in one of his great achievements, the unity of space-time and gravity (Albert Einstein, 1879–1955). The project still continues in the attempt to unite the standard model of physics with that of the strange quantum behavior of the very small.

Both *anti-realists* and *realists* can agree that mathematics provides the continuity. Despite the example of the quantum image of reality differing from that of our classical Newtonian picture of the world, "there is a smooth transition from the mathematical equations of quantum physics to the mathematical equations of classical physics."[26]

23. Coyne and Heller, *Comprehensible Universe*, 14.

24. Parsons's *It Started with Copernicus* is among the many works discussing these issues. See particularly ch. 6, "Mysteries of Method," which also includes a list of further readings. See also ch. 5, "Truth or Consequences."

25. It had been a staple of ancient thought, including that of Aristotle, that the heavens moved in perfect spheres while movement on earth was corrupted. It was the genius of Johannes Kepler to uncover another form of beautiful celestial motion—namely, the ellipse. It was then part of Newton's genius, together with that of the often-forgotten Robert Hooke, to understand these movements through a small number of equations that unified them. For an interesting discussion of Newton and what he owed to earlier students of the heavens, including his contemporary Robert Hooke, see Gribbin and Gribbin, *Out of the Shadow*.

26. Coyne and Heller, *Comprehensible Universe*, 141.

Since mathematics describes structures and since there appears to be continuity in *mathematical* structures between the images of the physical world as proposed by early and late theories of physics—the Newtonian world versus the quantum one, for example—then it follows that

> if mathematics is a "science of structures and their morphology," and if the essence of physics consists in applying these structures to investigate the world, an unavoidable conclusion is that successive physical theories disclose, step by step, the structure of the world. . . . Since this structure is a system of various substructures, one could say, analogously, that physics is a "science of the world structure and its morphology."[27]

One problem that immediately arises is the following: If mathematics does provide the continuity between different theories that is claimed for it and does therefore represent to us the fundamental patterns and structures of the world, how are we to account for the observation that any given physical theory may admit of two, or several, mathematical formulations? For instance, the "wave formulation" of quantum theory created by E. Schrödinger and the matrix formulation of Heisenberg are two different mathematical "pictures" of the same theory. There is a way, however, of conceiving this issue that accommodates it.

Ladyman outlines a solution to this problem: "We have various representations which may be transformed or translated into one another, and then we have the invariant state under such transformations which represents the objective state of affairs."[28]

In fact, as I shall later exemplify, at least one physicist has thought the discriminating factor between such a choice is the aesthetic properties manifested by the mathematical options.

Ladyman's conclusion is supported by Jill North, who offers an able assaying of what the relation of the "stuff" of physics is to the mathematics.[29] I quote sufficient of her description to orient the reader:

> Structure comprises the objective, fundamental, intrinsic features, the one that remain the same regardless of . . . conventional features of description. . . . We can learn about structure by looking at the invariant quantities. It is part of the structure of the world, according to a theory of physics, that it contain at least the amount of stuff needed to state the theory in an objective, coordinate-independent way—in a way best suited to

27. Coyne and Heller, *Comprehensible Universe*, 141.
28. Ladyman, "What Is Structural Realism?," 409.
29. North, "'Structure' of Physics."

> representing the nature of the world apart from our descriptions of it.[30]

Given the apparent independence of mathematical calculation and proof from the world of empirical and experience dependent truth, an issue arises as what *makes* the application of mathematics to the natural world comprehensible. Another way to express this is to ask what the world must be like if it is to be describable by mathematical physics. The evidence, to be clear, overwhelmingly supports the idea that it is so. What, in the light of this, is the relation between mathematical description, "stuff" of the world, and aesthetics?

PART 2

The Structural Beauty of Mathematics and Aesthetic Evaluation

I begin by considering a way in which mathematics and the structures of physics may be related so as to unite them within a single account. The purpose of this is twofold. First, it is to provide a plausible account of this relation that allows the properties of the physical world to exhibit aesthetic qualities without compromising either the objectivity of scientific method or the causal exteriority of the sub-particulate domain. Second, it is to open that domain to two corresponding modes of appreciation. On the one hand, there is the personal and "subjective" experience of beauty; on the other, there are objective features of structure—features that are themselves measurable—such as symmetry, proportion, and the simplicity of part-whole relations. The suggestion, in other words, is that aesthetic experience and structural description may converge upon the same underlying reality.

An Aristotelian Conception of Mathematics and Structure

A contemporary mathematician and philosopher of mathematics, James Franklin has proposed an account of the relation between world, science, and mathematics that is particularly helpful in this regard. In *An Aristotelian Realist Philosophy of Mathematics*,[31] he describes mathematics as the "science of quantity and structure," which is in fact the subtitle of the work. His central claim is that mathematical properties are not merely formal devices

30. North, "'Structure' of Physics," 66.
31. Franklin, *Aristotelian Realist Philosophy*.

or human projections but are real properties of structures—no less real than features such as dimension or shape.[32] Franklin defines a structure as

> a set of abstract entities with relations between them. . . . A property S is largely structural if and only if, "proper parts of particulars having S have some properties T . . . not identical to S, and this state of affairs is, at least in part, constitutive of S."[33]

Franklin's point may be clarified by considering a tartan pattern. What makes a piece of cloth a tartan is not merely the presence of particular colors but the ordered relations between them: stripes repeating at regular intervals, crossing symmetrically, and standing in fixed proportional arrangements to one another. The pattern therefore depends not simply upon the material particulars—the red, green, or blue threads—but upon the structural relations organizing them. Yet because the tartan still depends upon specific sensible features such as color, texture, and material embodiment, it does not yet constitute a purely mathematical structure.

Franklin's point is that mathematics abstracts from these contingent sensible properties and concerns itself only with the relational form capable of exact structural articulation. Mathematical description therefore does not attach arbitrarily to the world; it presupposes that the world itself exhibits structures whose relations are apt for mathematical representation.

This requirement is expressed more precisely in his definition of a purely structural property:

> A property is purely structural if it can be defined wholly in terms of the concepts same and different, and part and whole (along with purely logical concepts). In short, a purely structural property is one that can be defined in terms of Mereology and Logic.[34]

32. Franklin recognizes difficulties for his position that require response, including the problem of "uninstantiated universals" described by mathematics—that is, structures which do not physically exist. Another issue concerns "infinity," since it is by no means clear how infinity stands in relation to physical instantiation. He addresses some of these problems in Franklin, "Uninstantiated Properties."

Resnik is another philosopher known for defending the position that mathematics has a factual subject matter. He describes mathematics as the "science of patterns." In doing so, he also commits himself to a structuralist philosophy of mathematics and argues for mathematical realism, the view that the patterned structures described by mathematics genuinely exist. See Resnik, *Mathematics*.

33. Franklin, *Aristotelian Realist Philosophy*, 56.

34. Franklin, *Aristotelian Realist Philosophy*, 57.

If this is correct, then any object capable of mathematical description—whether a physical entity or an artifact—must possess structural properties of this kind. And since the entities studied by physics are preeminently those that admit of mathematical formulation, it follows that the physical world itself must be structured in ways that satisfy these conditions. The implication is significant: mathematical describability is not an external imposition upon the world, but a reflection of the kind of structure the world must already have in order to be describable at all.

This claim is, of course, open to debate, as are all attempts to unite mathematical entities with the "real" world. For present purposes, however, its importance lies less in its definitive truth than in the possibility it opens: that there may be a deep metaphysical unity between structure, mathematics, and—crucially—those features of structure that invite aesthetic evaluation.

It is perhaps worth a brief note that Franklin's definition of structure itself is not especially controversial, or idiosyncratic. For example, the physicist and cosmologist Max Tegmark, though differing in his metaphysical commitments, offers a closely related account. Tegmark is a realist about structure but adopts a Platonist position, holding that mathematical entities exist independently of the physical world they describe. Mathematical structures, he asserts, exist independently of the symbols we use to identify them.

Another representative of the independence of the ontology of mathematics is the Oxford mathematical physicist and Nobel laureate Roger Penrose, who defends a Platonist view:

> My sympathies lie with the Platonist view that mathematical truth is absolute, external, and eternal, and not based on man-made criteria; and that mathematical objects have a timeless existence of their own, not dependent on human society nor on particular physical objects.[35]

At the opposite pole are positions that treat mathematics as the outcome of human conceptual engagement with the world as a constructed framework, a human creation, that enables a certain kind of functional interaction with reality. This is broadly the "constructivist" or nominalist tendency.[36]

35. Lawrence and McCartney, *Mathematicians and Their Gods*, 5.

36. Marcus du Sautoy, professor of mathematics at the University of Oxford, has attempted to unite constructivist and realist positions in the following way. Structures describable in mathematical language, he argues, exist independently of us; this is the realist component. The mathematician, on the other hand, selects which among these

Franklin's position, as the title of his work suggests, is Aristotelian in spirit. It seeks to bring together the universality of abstract concepts with the particular objects that instantiate them. In doing so, it attempts to chart a course between Platonism and nominalism. Platonism is accommodated insofar as mathematics operates with universals, equations, relations, and entities that have general applicability. Nominalism is addressed by the claim that such universals are not free-floating but are embodied in particulars—in this case, in the mathematical properties of structures that make those structures what they are.

No doubt this position is as controversial and liable to be contested as any other in the philosophy of mathematics. Yet it has an intuitive appeal. It accords with the sense that any structure incapable of being captured by some appropriate mathematical formulation would thereby be excluded from realization in any possible world. The diagram below provides a simple but instructive illustration of this principle. The figure presents an apparently coherent object whose parts, when traced through their relations, are found to be arranged in ways that violate the constraints of consistent geometry. What appears at first as a unified structure dissolves under analysis into incompatibility. Such "impossible objects" make visible a fundamental point: not every conceivable arrangement of parts constitutes a possible structure, and where structural relations cannot be sustained within a coherent mathematical description, no corresponding object can exist.

are to count as mathematical structures; this is the constructivist component. The position may nevertheless be misleading, depending on precisely what is meant, since all possible structures presumably admit of mathematical description. See du Sautoy, "Exploring the Mathematical Library."

Example of an impossible structure.
The configuration shown cannot exist in three-dimensional space because its parts are related in ways that violate the constraints of consistent geometry. Such 'impossible objects'—popularised by M. C. Escher—illustrate how not every conceivable structure is mathematically or physically realisable.

What such images reveal is not merely a perceptual curiosity but a deeper constraint: that not all imaginable arrangements are structurally possible. And if structural possibility is governed by relations of the kind mathematics describes, then mathematics is not simply a descriptive tool but a condition on what can exist.

I cannot enter here into the full debate concerning the ontology and explanatory role of mathematics. Fortunately, my argument does not depend upon resolving it. What matters, and what is widely agreed across these differing positions, is that mathematics functions as the language in which the possible structures of the physical world are articulated. Whether those structures are discovered, constructed, or embodied, mathematics provides the medium through which their form—and, as I will argue, their aesthetic character—is made intelligible.

The Availability of Structures to Mathematical Description and Cognitive Representation

The force of Franklin's claim—that "mathematical entities can be literally exemplified in the world"[37]—is, for present purposes, clear and significant. If the properties described mathematically are not merely formal constructions but are instantiated in the structures they describe, then those structures are not only available to logical and mathematical reasoning. They are also, in principle, available to forms of cognitive apprehension that extend beyond formal deduction: to imaginative, abstract, and even quasi-visual modes of perception. In such cases, the mathematician or mathematical physicist does not simply calculate or infer but, in a more substantive sense, *grasps* structure—apprehending relations that are at once intelligible and, in a disciplined way, perceptually present.

This claim, though philosophical in origin, receives a degree of support from experimental work in the cognitive psychology of representation. Cooper and Shepard, for example, demonstrated that subjects presented with a complex three-dimensional structure were able to form and manipulate an internal representation of it.[38] Having first seen the object, subjects were then shown a series of photographs and asked to determine which depicted the same object viewed from a different orientation. The task required them not merely to recall the object but to *mentally transform* it, rotating it within an internal space in order to judge identity across differing perspectives.

37. Guaspari, "Sum of Its Parts," 63.

38. Cooper and Shepard, "Chronometric Studies."

Examples of images used in the Cooper and Shepard mental-rotation experiments

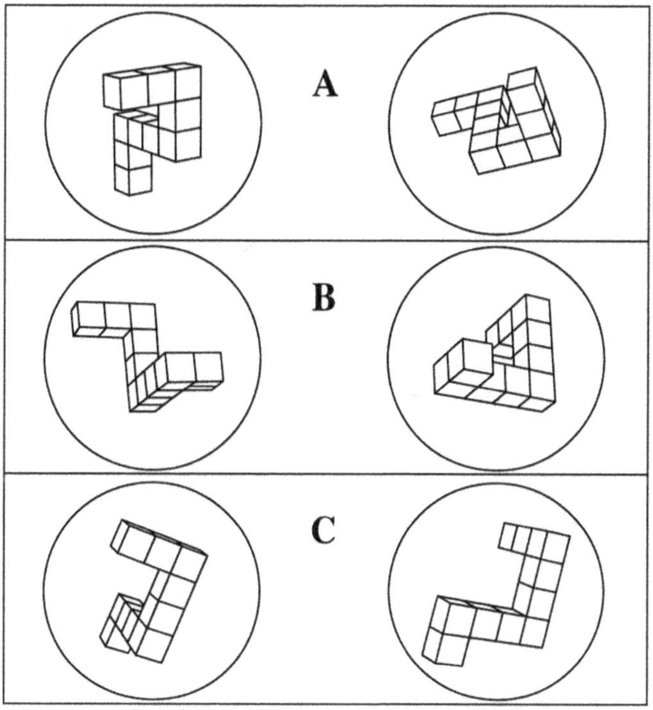

Can the two figures in each set be made identical by rotating them in space?

What this suggests is not simply a capacity for memory but a more sophisticated ability: the capacity to construct and operate upon structured internal representations. The subject is able to preserve the relational organization of the object while altering its orientation, and thereby to draw conclusions about identity and difference under transformation. Such findings indicate that structural relations, once apprehended, can be held in a form that is neither purely sensory nor purely propositional but intermediate, an abstract yet manipulable image.

If this is so, then the availability of structure to mathematical description is complemented by its availability to cognitive representation. The same structural features that render an object amenable to mathematical formulation also render it, at least in certain cases, accessible to imaginative and perceptual apprehension. This convergence is important. It suggests that the intelligibility disclosed in mathematical description is not wholly detached from human modes of understanding but can, under appropriate conditions, be *seen*—not in the ordinary sensory sense but in a disciplined, cognitively mediated way that makes structure available to both thought and perception.

Like Einstein or Kepler, a scientist may be fortunate to be blessed with a visual or abstract imagination that can conceive of different possible structures and judge them according to their conceivable or inconceivable plausibility. That imagination will coincide where it correctly discerns structural arrangements with potential mathematical descriptions of it. Among the structural features they may internally utilize to accomplish this end are aesthetic values.

An Epistemic Model of Dual Access to Structural Aesthetics

I want to develop this idea by a relatively simple example of how it might work. The purpose of what follows is to outline the general form of a credible appreciation of beauty in two ways that unites the "subjective" and the "objective" as epistemological tools that exploit different human faculties to the same scientific and ontological ends.

I am relying here on the understanding which I develop that the phenomenal experience of beauty can yield objective truths about aesthetic properties. Our "subjective" responses, in other words, contrary to common opinion, can carry truth-evaluable content about the aesthetically expressive properties of a structure.[39]

I start by considering how, if a person wishes to create something of beauty, something of the relations of the structure being created and how to attain "beauty" using them must be understood.

It has long been known, for example, that there is a connection between beauty, proportion, and mathematical symmetry, known as the "golden mean." Architects, artists, and others have for many hundreds of years understood that a certain geometric ratio between the height and width of a rectangle is one which is most aesthetically satisfying.

39. Breitenbach's "Beauty in Proofs" offers a Kantian account of the experience of aesthetics in mathematical judgment. Her position is summarized in the abstract: "I show that on the Kantian account, beauty in mathematics is a non-conceptual response felt in the light of our own creativities involved in the process of mathematical reasoning."

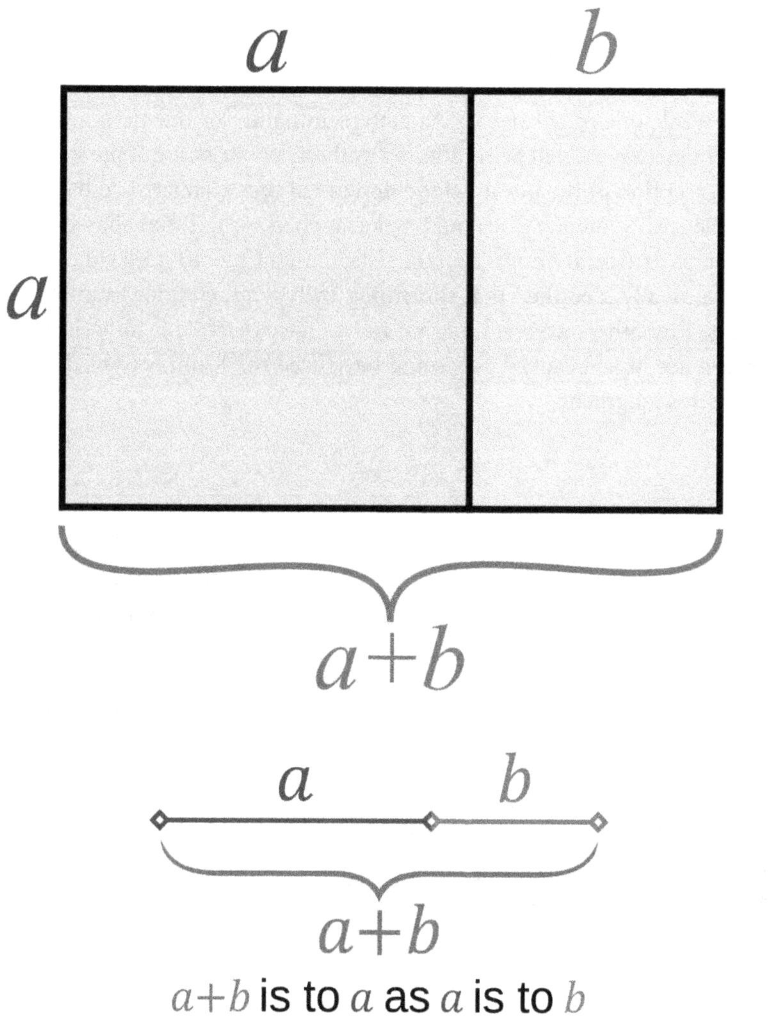

The golden ratio

Rectangles constructed in such a way that do not meet this ratio typically attract aesthetic epithets such as "too squat," "too narrow," and so on.

If an artist, architect, or anyone engaged in the production of an artifact wishes to produce for some purpose a work that is discordant, then his construction will accordingly violate the mathematical principle, but if he wishes to produce something that induces feelings of perceptual satisfaction and exclamations of its beauty, he will obey the principle.

It does not follow, of course, that an artist or artisan be familiar with the mathematical ratio. All that is required is that he recognizes it as pleasing and works towards it.

While one might perceive a pattern in nature or one designed by an artist or artisan *without* being affected by the aesthetic shape of the structure, the perception of the beauty *of* the design *requires* qualitative and emotive experiences. Someone who could make an *inference* to beauty based on, for example, structural design characteristics, might be said to have correctly and rationally identified that something with symmetry, for example, was counted by others as beautiful, but not to have *detected* it himself via the experience of *its* beauty.[40] He would have used the features of its *design* to make this judgment.

Michelangelo's Sistine Chapel showing use of the golden ratio

Something about the beauty *in* the thing itself might be said to be left out in this case. Someone entirely incapable of experiencing the beauty of something might be thought to lack what is most essential to beauty

40. Someone bereft of the capacity for aesthetic responses may *not* be able to pick out something beautiful by experiencing it. Nevertheless, they may master other ways of detecting beauty. They may, for example, have learned that structures which obey the principle of the "golden mean," a ratio shared by some geometric structures, are generally those which most observers call "beautiful." Similarly, a mathematician, similarly aesthetically ill-disposed, may nevertheless recognize that certain features of equations—their simplicity but expansive explanatory properties, for example—may be those which her colleagues are inclined to call "beautiful." It also seems possible that those mathematicians equipped to make good aesthetic distinctions might use this ability to increase the brevity of mathematical decision-making in certain circumstances.

itself—that is, its power to provoke the qualitative response I have discussed. If the beauty is real, then he also has been unable to access part of the reality of the structure he is observing.

It is thus perfectly possible for the mathematician or scientific theorist to technically exploit such features as "symmetry" or "simplicity" in making theoretical decisions without experiencing the full range of aesthetic responses that those features make available to cognition, perception, and emotion.

The congruity between the "structure," the description, and these responses means that for those who *are* suitably cognitively, educationally, and culturally equipped, such experiences *are* available for theoretical use. They are provided, as it were, with *two* theoretical ways of accessing the shape of a structure.

Structural Integrity, Truth, and Beauty

There is a common experience of the "rightness" of a particular structure that extends through reasoning, the physical, the abstract, and even the historical and the personal. "Harmony," "unity," "proportion," "symmetry," and groupings of concepts associated with the use of these words are common features assigned to positive aesthetic evaluations across a range of domains, despite the substantial variation of their use in the context of human affairs.

They are frequently exploited when someone is trying to offer some description of an object or event that has stimulated an experience of beauty. This is displayed in the example given above of someone describing a rectangle as "ugly" or "squat" which does not obey the golden ratio but who does not have mastery of the necessary equation that describes the relation precisely.

The Nobel physicist Werner Heisenberg (1901–1976) tried to explain the idea of the relation of aesthetic parts and whole in physics in an article titled, "The Meaning of Beauty in the Exact Sciences."[41] There, he defined it as the "proper conformity of parts to one another and to the whole." To explain what he means he cites Newtonian mechanics, with parts being the "individual mechanical processes" and the whole, "the unitary principle of form which [the] processes comply with, and . . . [which] was mathematically established by Newton."[42]

It is thus not a coincidence that traditional classical metaphysics incorporated in the experience of beauty a strong sense of the *rightness* of things,

41. See Heisenberg, *Across the Frontiers*.

42. Heisenberg, *Across the Frontiers*, 174.

itself indicating "truth." The best known of these is the Platonic order: to properly grasp the true nature of reality is to become acquainted with the order of abstract universals at the apogee of which sits the form of goodness, truth, and beauty ordering the universals beneath it into a perfectly harmonious structure, each occupying its appropriate place.

Platonism is not a necessary consequence of the recognition of the sense of something being rightly *ordered*. Shorn of Platonist adornments, the idea of an "ordered" structure offers a different conception of how a theory might fit together, and its representational capacities "mirror" effectively the structures it attempts to articulate.

Here, aesthetics may have a particular role to play; the greater the aesthetic merit of a given structure, the greater the sense that its parts are maximally ordered. Indeed, maximal order might be coincident with the greatest beauty available to it to manifest.

Henri Poincaré (1854–1912), the French mathematician and theoretical physicist, tried to articulate the nature of beauty in his discipline in this kind of way:

> I am not speaking . . . of the beauty of qualities and appearances. . . . What I mean is that more intimate beauty which comes from the harmonious order of the parts, and which intelligence can grasp.[43]

He proposes that the purpose of science is to

> offer an understanding of the relations between phenomena. . . . Scientific theories offer a . . . manner of bringing together facts which appearances separate. . . . [It is in that] harmony that our theories find beauty.[44]

Nicholas Maxwell's argument concerning simplicity suggests a role for aesthetic judgments when faced with a choice between different theories capable of accounting for the same phenomena. Why should simplicity matter? Maxwell argues that it is always possible to construct indefinitely many alternative theories that agree with an accepted theory over observed phenomena while differing with respect to unobserved cases. As a result, empirical evidence alone can never determine a unique theory. If observation were the sole criterion governing theory choice, it would be difficult to explain why such rival theories are routinely rejected.

43. Ivanova, "Poincaré's Aesthetics of Science," 2585.

44. Ivanova, "Poincaré's Aesthetics of Science," 2585.

Maxwell maintains that scientists therefore appeal to additional criteria such as simplicity, unity, explanatory depth, and coherence when evaluating theories. These standards operate alongside empirical adequacy and help distinguish between competing accounts of the same phenomena. In relying upon such principles, science implicitly assumes that the world possesses an underlying order corresponding to them. For Maxwell, this poses a challenge to traditional empiricism, since the preference for simple and unified theories cannot itself be justified solely by appeal to observation.

What holds true for simplicity or parsimony, as some theorists prefer to call the principle, can also hold true for aesthetic structural features, with suitable modifications to reflect their distinctive properties, and perhaps even for beauty itself as the phenomenal outcome of an experience of the maximal order of a structure.[45]

The aesthetic judgments associated with all this may be in the end an attainment of *normative* intuition[46] rather than the end of a chain of reasoning itself.[47] This does not assign the possibility of accurate perception and judgment of aesthetic order and its component structures to the variability of subjective experience without further ado.[48]

The axiomatic foundations of mathematical proofs and formal logic, for example, rest ultimately on the grasping of certain initial truths. A teacher can only present examples and say, "This is how you ought to reason if reasoning is what you want to do." There are no further principles to grasp that can illuminate further what reasoning in this context is. At some point, the student will simply have to see "This is how one goes on,"[49] and manifest

45. Maxwell, *Comprehensibility*. I discuss structure, objectivity, and subjectivity and their relation to "beauty" in a later chapter considering what it is we mean when we talk about "beauty."

46. I am not making the mistake of suggesting that logical relations are created by us. I am simply pointing out that whatever *accounts* for the necessity, the "hardness," of logic, we have to grasp its fundaments *before* we can deduce anything.

47. The truth of the propositions should not be confused with the psychology of coming to *see* that they are true.

48. Engler argues that the aesthetic experience of the scientist is both "perceptual-intellectual" and "sensuous-emotional," "which it is very hard to differentiate." Engler, "Aesthetics in Science and Art," 31. He is less concerned with the contribution of this experience to theory evaluation than with the nature of the experience itself, a question that remains an interesting one if such precision is attainable.

49. Wittgenstein argued that there is no "fundamental meaning" which a student must grasp in order to learn what it is to "follow a rule," whether in mathematics, language, or logic. In simplified form, his position was that there is no magical "semantic internal event" switched on by the "right" explanation that finally illuminates the meaning of a rule. Rather, a student's grasp of a rule is exhibited in the ability to use it correctly. Wittgenstein offered numerous examples to illustrate this view. Substantial

it in their dealings with the principles in question. The point of raising this example is not to draw an exact analogy with the position I am arguing for but simply to illustrate that at some point in any learning, the possibility of giving further instructions will have to end. The student will have to grasp what is at issue, and that grasp will be manifested in their capacity to demonstrate that in how they accomplish the task.

To experience the beauty of a "deep theory" of physics, for example, is perhaps to grasp or perceive something of the rightness *of an order of things* in response to the properties of its structure. At some point the only answer that the percipient physicist can give is, "This is the right order of things." The weight to be assigned to judgments exploiting one or another aesthetic quality will range across different variables according to discipline, subdiscipline, subject matter, its relations to other theoretical variables, the quality of the data and evidence, and so on.

When to accept that a theory is to be appropriately dismissed or accepted is a long and difficult question perhaps, but whether or not this rests ultimately with the community of scientists, the individual, or involves a period of turbulence and theoretical conflict, it seems that ultimate theory evaluation does not rest on any algorithmic procedure. Despite, therefore, the hardness of discerning real beauty and a reliance on qualitative assessment, like the use of other values in theory decision-making, this does not single "beauty" out as of no theoretical utility. Defenses of science, as in the case of Dawes, often point to the place of evidence. "Evidence," they say, is what distinguishes science. If one has a hypothesis and puts it to the test, the evidence will decide the issue regardless of the individual scientist's personal hopes or beliefs. A step back, however, shows us that we first have to grasp why evidence bears the truth-evaluable relation to theory that it does. This understanding is an achievement of culture and education and, as I have suggested earlier, is itself the outcome of a developed view of what external nature is and how to learn about its causal functioning.

The accomplished mathematical physicist brings his range of aesthetic appreciation to his work, in which the nature, order, and aesthetics of the structure "leap out" at him. It is precisely here that we see the conceptual uniting of the metaphysical structure, the empirical nature of the aesthetic property, and the availability of normative judgments as theoretical tools for the practicing scientist. This is a product of a long practical and intellectual apprenticeship. He becomes, one supposes, like the art critic who can place

---

literature has arisen around these issues, bearing upon fundamental questions in cognitive psychology and neuroscience. Useful and opinionated summaries are provided by Rundle, "Meaning and Understanding," and Arrington, "Following a Rule."

a style of painting immediately within one school, time, or culture.[50] Attaining such a status, of course, never results in entirely harmonious agreement between those of similar status, neither might they do so in the case of aesthetic judgments.

Aesthetic Judgments in Mathematical Physics: Case Examples and Discussion

Is it a problem that a particular skill which relies on subjective honing rather than objective measurement to detect an aesthetic property is required in a scientific context? Not at all. As I made clear, the correctness or otherwise of such an attribution is dependent for its confirmation on the "actual" structure as eventually revealed by the evidence. The judgment is therefore not detached from empirical reality but answerable to it. The role of aesthetic discrimination in science is not to replace evidence but to assist in the recognition of forms whose full significance may only later become apparent.

I want to exemplify these claims in practice. The first example I present discusses a property that lies at the heart of fundamental physics: the feature of "symmetry." It is important to distinguish two things here. Firstly, it may be the case that symmetry is a feature of structures that tends to elicit an experience of beauty, but it is not itself identical to beauty. It also illustrates the epistemological congruence I described in the golden ratio explication. The second example discusses the mathematical physicist P. A. M. Dirac (1902–1984), who made the explicit claim that the beauty of a theory was fundamental to its truth. As I shall show, Dirac stressed not merely isolated mathematical features but the beauty of a theory apprehended as a whole. Such judgments suggest a profound familiarity with the structures governing sub-particle and other physical processes, one only available through extensive immersion in the mathematics and theories of the discipline.

Before turning to specific examples from mathematical physics, it is important to appreciate the extraordinary scope of the idea of symmetry itself. In ordinary language symmetry is often associated with simple forms of balance or repetition: the equal sides of a square, the repeated pattern of tiles on a floor, or the way a melody may retain its identity when played in a different key. In physics and philosophy, however, the significance of symmetry is much deeper. Symmetry concerns the persistence of intelligible relations through change. Something counts as the same object, structure, or law only because certain features remain invariant under alteration of time, place, orientation, or circumstance.

50. Steane, *Science and Humanity*, 27–28.

This requirement reaches far beyond physics alone. Ordinary experience already depends upon it. A chair remains recognizable as the same chair whether it is moved from one side of a room to another, viewed from a different angle, or encountered at different moments of the day. Likewise, language itself depends upon stable relations: words must preserve enough continuity of meaning through repeated use for communication to occur at all. If absolutely everything altered entirely and randomly from one instant to the next, there would be no stable objects, no identifiable events, and no coherent world available for recognition.

Even the ordinary notion of causation depends upon ordered relations of this kind. We ordinarily distinguish between a sequence in which one event gives rise to another and a random succession of disconnected happenings. If striking a glass causes it to shatter, we do not ordinarily suppose that the shattering caused the striking. Cause and effect therefore depend upon stable asymmetrical relations whose intelligibility presupposes an underlying continuity and invariance in the structures involved. A world lacking such ordered relations would not merely be difficult to investigate scientifically; it would be difficult to recognize as a world at all.

Symmetry in this broader sense therefore concerns the conditions under which structures, identities, laws, and causal relations can exist as stable realities. The importance of the concept in modern physics lies precisely in the fact that the mathematical laws describing nature repeatedly display forms of invariance that make coherent physical explanation possible. Symmetry is thus not merely an ornamental feature added to already existing structures, but appears deeply connected to the intelligibility of the structures themselves.

The first example is drawn from a work by the quantum physicist Andrew Steane of Oxford University. In his 2018 book, *Science and Humanity: A Humane Philosophy of Science and Religion*,[51] he discusses and exemplifies what are, in essence, two modes of access to the properties of structures that I have just discussed. He makes his example usefully accessible to the non-physicist and wishes to "explore how it can be that different areas of scientific discourse do not replace one another but still connect with one another."[52]

This is not, he says, merely a question about scientific discourse itself but about the nature of the physical world, one arising directly from the area of science which is his own field of expertise: fundamental physics. The

51. Steane, *Science and Humanity*.

52. Steane, *Science and Humanity*, 23.

issue with which he is concerned is the relation of symmetry to physics and to the equations describing forces.

One of the fundamental features prominently associated with the successes of modern physics is the importance of the principle of symmetry. Symmetry, of course, is fundamental not only to physics but to all forms of physical and abstract structure. Every and any structure can be judged by its symmetrical form in one way or another. In philosophical terms, the existence of any structure is logically dependent upon some form of symmetry relation. It is also fundamental to classical conceptions of beauty and to the arts generally. Steane argues that principles of symmetry "when brought to bear with mathematics yields rich and powerful insights into many physical phenomena."[53] The *symmetry principle* states that "certain types of change should not make any difference to what is going on in some physical behaviour."[54]

There are many forms of technical symmetry. An object such as a circle may be said to possess rotational symmetry because its shape and relations do not change when rotated about its axis. A crystalline solid which remains unchanged under mirror reflection is said to possess mirror symmetry. Steane then moves from visible structures to more abstract ones. One can, he points out, have symmetries not only of objects but also of equations and laws of motion. An equation describing particle collisions, for example, may possess mirror symmetry if reflected trajectories leave the equation unchanged. Likewise, equations may possess translational symmetry if the behavior they describe is unaffected by moving the system from one place to another.

53. Steane, *Science and Humanity*, 24.

54. Steane, *Science and Humanity*, 23.

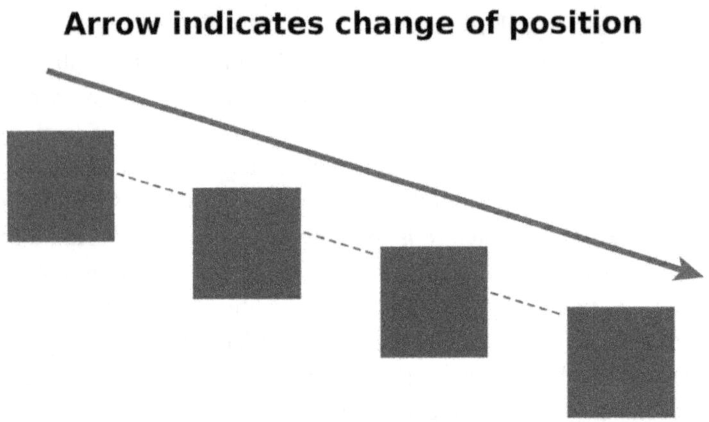

An example of translational symmetry:
one square moved through four different positions

His own example concerns this principle of translational invariance.[55] A car engine functions in essentially the same way regardless of where the car happens to be. If one drives around China, as he says, the engine is not affected by the car's distance from the Eiffel Tower. Its functioning depends upon local conditions such as road gradient, pressure, or air temperature, not upon its absolute location in space. The operation of the engine therefore remains invariant when translated from one place to another. At first sight this may appear philosophically trivial, and Steane therefore imagines the response of a practical engineer who might object that in analyzing the functioning of the engine one need only employ Newton's laws of motion and the chemistry of combustion. The engineer may never explicitly invoke translational invariance at all. On this view, symmetry appears irrelevant to the explanatory work actually being done. The equations already describe the sequence of causes and effects; symmetry merely appears as a property the equations happen to possess.

The objection is important because it captures a common modern assumption: that symmetry is secondary to the physical laws themselves, an elegant feature perhaps, but not something fundamentally important. Steane's reply shifts the issue from prediction to intelligibility itself. What the engineer has failed to grasp, he argues, is that "symmetry principles

55. Steane, *Science and Humanity*, 24–25.

contribute before the equations themselves are discovered because they place conditions upon what sorts of equations can make sense."[56]

As soon as one supposes that the engine can be described "in and of itself," translational invariance has already been assumed. Even the possibility of a general equation describing the engine requires the prior assumption of symmetry. Symmetry is therefore not an afterthought arising from microscopic causes nor merely "a pleasing or helpful thought." It belongs to the intelligibility of physical law itself.

Symmetry principles therefore function as what he calls "meta-laws," higher-order principles constraining the form acceptable physical laws may take. They are not merely convenient heuristic devices nor simply pleasing regularities retrospectively noticed in successful theories. Rather, they express something deeper about the intelligibility of the structures themselves. Steane makes this point explicit when he says that symmetry is not merely useful because it helps us formulate correct physical models but because "it gives us deeper insight into what we find."[57] The physicist and theist Stephen M. Barr concurs and points out that every one of the four fundamental forces of nature—gravity, electromagnetism, the strong nuclear force, and the weak nuclear force—is based upon principles of symmetry.

> [To] say that these forces are "based" on symmetries, we mean several things . . . profoundly, the very fact that there are such forces in nature is a consequence of these symmetries . . . the characteristics of these forces are controlled by their symmetries . . . *the structures of the mathematical laws governing these forces is to a large extent determined by their underlying symmetries.*[58]

This is a position supported also by the Oxford mathematician Marcus du Sautoy, who remarks that "the symmetries of a structure often reveal secrets about how it behaves."[59] What emerges from these examples is that symmetry is not merely decorative or psychologically satisfying. It belongs to the structure of intelligibility itself. The aesthetic and the intelligible are internally related.

My second example is that of Paul Dirac, Nobel Prize–winning physicist and one of the principal architects of modern quantum theory. Dirac

56. Steane, *Science and Humanity*, 24–25.

57. Steane, *Science and Humanity*, 24–25.

58. Barr, *Modern Physics and Ancient Faith*, 96; emphasis added.

59. Du Sautoy, "Grand Designs," 92. Du Sautoy gives examples from disciplines other than physics. Symmetry, he notes, is central to the classification of possible crystal structures in chemistry, while in biology the functioning of a virus depends significantly upon its shape.

shared the 1933 Nobel Prize for "his discovery of new fertile forms of the theory of atoms and for its application."[60] His work was driven by the desire to construct an equation capable of describing the behavior of elementary particles in a manner consistent both with Einstein's theory of relativity and with the strange and counterintuitive phenomena emerging from quantum mechanics.

The resulting equation proved revolutionary. It combined relativity and quantum mechanics within a single mathematical structure, provided a natural account of electron spin, and led directly to Dirac's prediction of antimatter. As the historian and philosopher of science Helge Kragh notes, Dirac "often stressed that predictive ability is not enough for a physical. He increasingly turned to the view that scientific understanding must include criteria such as beauty."[61]

Dirac himself was explicit about his commitment to aesthetic value and believed that Einstein had been guided by similar principles. He wrote of Einstein's work, "We owe it to Einstein . . . that one needs to have beauty in mathematical equations which describe fundamental physical theories."[62]

Discussing Einstein's development of general relativity, Dirac argued that Einstein "was *not* trying to account for some results of observations. Far from it. His entire procedure was to search for a beautiful theory."[63]

Einstein's achievement, in Dirac's view, lay not merely in explanatory success but in the extraordinary "simplicity and elegance"[64] of the mathematical structure itself. Somehow Einstein had grasped the possibility of connecting gravitation with the curvature of space and had developed a mathematical scheme capable of expressing the idea coherently. The beauty of the equations was not decorative but internal to the intelligibility of the theory itself.

Dirac's position becomes clearer still in his reflections upon the development of quantum mechanics during the 1920s. Niels Bohr had proposed that electrons moved around the nucleus in fixed orbits, occasionally "jumping" from one orbit to another. As Dirac remarked, "We could not picture how the jump took place. We just had to accept it as a kind of discontinuity."[65]

60. Kragh, "Seeking Beauty," 28.
61. Kragh, "Seeking Beauty," 28.
62. Dirac, "Einstein Theory," 6.
63. Quoted in McAllister, "Dirac and the Aesthetic Evaluation," 91.
64. Quoted in McAllister, "Dirac and the Aesthetic Evaluation," 91.
65. Dirac, "Physicists' Picture of Nature," 46.

Two major approaches emerged in response to these problems. Werner Heisenberg adhered closely to experimental evidence concerning the spectral lines associated with atomic transitions and discovered that the data could be organized through an existing mathematical procedure known as matrix mechanics. Erwin Schrödinger, by contrast, pursued a more overtly mathematical route, seeking what he described as a "beautiful theory for describing atomic events."[66] When Schrödinger first derived his wave equation, however, it failed properly to account for certain features of the hydrogen atom. The discrepancy distressed Schrödinger and delayed publication of the theory. Later developments resolved the difficulty, and Dirac drew from the episode one of his most famous conclusions:

> There is a moral to this story, namely that it is more important to have beauty in one's equations than have them fit experiment.[67]

The point was not that experiment is dispensable nor that physicists may simply ignore empirical evidence. Dirac's claim was subtler than that. Mathematical beauty, he believed, may disclose a structural truth not yet fully articulated empirically. Temporary discrepancies may arise because a theory is incomplete in its development rather than fundamentally mistaken. As he continued, "If one is working from the point of view of getting beauty into one's equations, and if one really has a sound insight, one is on a sure line of progress."[68] Even where agreement with experiment remained imperfect, "the discrepancy may well be due to minor features that will get cleared up with developments in the theory."[69]

It is important not to misunderstand the role beauty played in Dirac's thought. Beauty was not, for him, a decorative accompaniment to successful physics nor merely an emotional response to equations already known to be true. Rather, aesthetic judgment functioned as a methodological and epistemic principle internal to theory construction itself. Dirac repeatedly trusted mathematically elegant structures in advance of empirical confirmation, believing that sufficiently deep harmony within the mathematics disclosed something real about the structure of nature. The remarkable historical fact is not simply that Dirac held such views but that this procedure proved extraordinarily fertile scientifically.

66. Dirac, "Physicists' Picture of Nature," 47 (Dirac quotes Schrödinger as saying this to him).

67. Dirac, "Physicists' Picture of Nature," 47.

68. Dirac, "Physicists' Picture of Nature," 47.

69. Dirac, "Physicists' Picture of Nature," 47.

After 1935 Dirac's conviction concerning mathematical beauty deepened still further. Kragh notes that he increasingly came to believe that the laws of nature themselves are "characterised by a high degree of mathematical beauty." Dirac wrote in 1939, "As time goes on . . . it becomes increasingly evident that the rules which the mathematicians have chosen are the same as those which nature has chosen."[70]

An important issue for present purposes is what Dirac thought mathematical beauty actually was. As Kragh points out, Dirac never properly defined the concept nor did he think it necessary to do so. Mathematical beauty is, he maintained, "a quality which cannot be defined, but which people who study mathematics have no difficulty in appreciating."[71]

One reason for Dirac's conviction may lie in the distinction he drew between artistic beauty and mathematical beauty. Mathematical beauty, he claimed, is the same in every country and in any period of time.

According to Dirac, mathematical beauty possesses an objectivity lacking in merely local taste or cultural preference. Jane McDonnell similarly argues that mathematics displays a timeless beauty associated with intelligibility, coherence, symmetry, simplicity, generality, and truth. This need not imply that all future physical theories will instantiate these qualities in identical ways. Rather, the claim is that mathematical structures possess an objective intelligibility whose aesthetic character is not reducible to subjective preference, even though the particular forms through which such intelligibility is realized may vary historically and conceptually.

Kragh, however, does not take Dirac's view at face value. His criticism is that beauty is subjective and therefore cannot function as a reliable guide in science. Rather than theories being accepted because they are beautiful, he argues, they come to be regarded as beautiful because they are empirically successful. This objection, however, rests upon precisely the dualistic metaphysical picture challenged earlier in this work. It may be unfair to Kragh, but one cannot help the thought that that dualism is so deeply ingrained in modern thought that it has come to seem something of an alethic truism that aesthetic apprehension belongs entirely to subjective projection rather than to the disclosure of real structural properties.

Yet it is entirely possible, contra Kragh's view, that empirical success stimulates perceptions of beauty, that empirical success may deepen recognition of a beauty already genuinely present but not initially fully perceived. Explanatory scope, simplicity, elegance, coherence, and unifying power may all become clearer retrospectively. That retrospective recognition does not

70. Kragh, "Seeking Beauty," 30.
71. Kragh, "Seeking Beauty," 31.

invalidate the reality of the property recognized. It is a commonplace of human experience after all, that growing understanding of anything—a person, a painting, a work of literature, even a theory—can provoke a greater appreciation of its intrinsic qualities, including those of the aesthetic. Indeed, a growing understanding of, say, a particular work of art may lead to a generalized understanding of other similar works, precisely because one has learned to recognize those properties elsewhere. Many truths about a theory may emerge only gradually through its successful application. It may, for example, reveal a wider explanatory range than was initially apparent or disclose deeper unities between previously disconnected phenomena. It is not inconceivable that beauty itself may become more visible through precisely such developments.

Nor does Kragh provide any convincing explanation as to why Dirac continued assigning such epistemic significance to beauty despite periods of empirical uncertainty. Dirac was almost wholly absorbed in mathematical physics. It seems unlikely that he attributed such importance to beauty casually or sentimentally. His position remained unequivocal: "There are occasions when mathematical beauty should take priority over agreement with experiment." And again, in 1980, "The foundations of a theory are, I believe, stronger than what one could get simply from the support of experimental evidence. The real foundations come from the great beauty of the theory."[72]

J. McAllister summarizes Dirac's position succinctly. If the aesthetic features of a theory warrant sufficient confidence, then unfavorable empirical findings should lead one first to question the auxiliary assumptions surrounding the theory rather than immediately abandon the theory itself. Dirac himself expressed the matter forcefully:

> Suppose a discrepancy . . . appeared, well confirmed between the theory and observations. . . . Should one then consider the theory wrong? The answer to the last question is emphatically no. Anyone who appreciates the fundamental harmony connecting the way nature runs and general mathematical principles must feel that a theory with the beauty and elegance of Einstein's theory (for Example), *has* to be substantially correct.[73]

Taken together, these examples suggest that aesthetic judgment in mathematical physics is neither peripheral nor dispensable. Symmetry principles constrain what can count as intelligible physical law while

72. Dirac, "Physicists' Picture of Nature," 47–48.

73. Quoted in McAllister, "Dirac and the Aesthetic Evaluation," 95; emphasis Dirac's.

mathematical beauty may guide physicists towards theories whose explanatory coherence exceeds what is immediately available empirically. Far from standing opposed to objectivity, aesthetic judgment may therefore belong to the very conditions under which the structures of the world become intelligible at all.

Dirac's Equation

I want to end this section with a highly simplified version of the Dirac equation, indicating how each of the factors mentioned by McDonnell are on display.

Dirac's work in 1927 was motivated by concerns about the incompatibility of Einstein's special theory of relativity with contemporary versions of quantum mechanics. At its most elementary, the problem was to describe the motion of the electron and the atom in a way consistent with relativity, since the Newtonian model broke down in describing the behavior of micro-phenomena. What was required was the incorporation of space-time as an essential factor in the description. Dirac set about trying to find a mathematical equation that described the behavior of a particle consistent with both theories.

I include that equation below. It is not necessary, however, to my argument that one comprehend the mathematics in any technical sense. What matters is that the reader can recognize, even at a distance, that a small set of symbols and functions is able to capture, in a unified form, an extraordinarily complex range of physical behavior. The aesthetic point begins here: that a vast domain of phenomena can be brought under a single, tightly ordered expression.

In order to make this minimally accessible, the version I display is owed to Graham Farmelo, fellow of Churchill College, Cambridge, and Dirac's biographer in *The Universe Speaks in Numbers.*

$$(p_0-\alpha_1 p_1-\alpha_2 p_2-\alpha_3 p_3-\alpha_4 mc)\psi=0$$

What the symbols denote:

1. P_{0} denotes the energy of an electron with mass m.
2. The remaining P symbols represent the momentum of the particle in three spatial dimensions.
3. The α symbols represent the "Dirac matrices," each consisting of a square array of sixteen numbers.

4. \(\Psi\) encodes a mathematical object known as a "spinor," which describes the behavior of the electron.

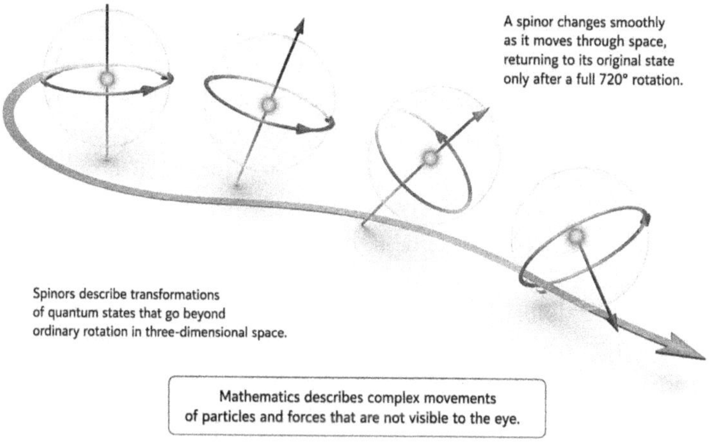

A visual representation of a spinor

A simple way of grasping the image of a spinor is to envisage it as a spinning top being rotated through 360 degrees. A highly simplified way of relating it to Dirac's equation is to see the "top" as the particle and the circles as the generation of electromagnetism.

With his equation Dirac explained why an electron has a spin and why it has an associated magnetic field, thus uniting Maxwell's electromagnetic waves with the quantum behavior of particles. Later, Dirac predicted from the mathematics alone the existence of an anti-electron with the same mass but an opposite electric charge—a prediction subsequently confirmed experimentally.

Each of McDonnell's criteria—"timeless beauty linked to ideas of simplicity, symmetry, generality and truth"[74]—is instantiated here by the equation. To see how this is so, it is helpful to make explicit what these terms amount to in this context.

The simplicity in question is not the simplicity of ease but of compression: the equation gathers together, in a remarkably economical form, relations that would otherwise require many separate descriptions. A small number of symbols carry a large explanatory burden. The symmetry lies in

74. McDonnell, *Pythagorean World*, 19.

the way the equation treats space and time, and the different components of motion, under a single formal scheme: no direction or dimension is privileged, and the same structural relations hold throughout. Its generality is shown in the range of phenomena it governs—any electron, in any context consistent with the theory, falls under its description. And its truth is not merely empirical adequacy, but the capacity of the equation to generate new, previously unobserved phenomena, as in the case of the positron.

In order to make these features more familiar, we can step back briefly to consider how mathematics in general exemplifies them. Mathematics presents us with a rigorous language whose various manifestations are underpinned by logical precision and exact rules of manipulation and deduction. The truths that proceed from mathematical reasoning are valid wherever the same relations obtain; they do not depend on time, place, or perspective.

Consider a simple example: $10/2\ (=5) > 6/3\ (=2)$. This expresses a relation that holds necessarily wherever the same structure is instantiated. Its "beauty," in however minimal a sense, lies in this combination of clarity, necessity, and invariance. It is symmetrical in that it holds under any instantiation of the same relations; general in that it applies universally; and true in virtue of its logical form. It is perhaps worth noting that this bears some similarity to the beauty of a poetic aphorism that captures profound truths in a simple sentence whose depth and range are immediately resonant with the sensitive reader. We are so familiar with syntax, however, as the organized bearer of our meanings that we lose sight of the extraordinary fact that sounds (or written shapes) somehow manage to carry with their brute physicality a fecundity of meaning for us and, sometimes, a painfully exquisite existential beauty. For those who can grasp the similarity, mathematics is equally fecund in the same aesthetic dimension.

Put more generally, many mathematical expressions possess the features I have gestured to in the "simple" example, and in that sense may be said to be beautiful. As I have suggested, all natural structures manifest mathematical form as intrinsic to their being instances of natural items at all, though not all mathematically possible structures are realized in the physical world.

When we turn to Dirac's equation, however, something occurs in the sense I discussed above when referring to the poetic aphorism. First, highly abstract mathematical structures—matrices, spinors, and relativistic relations—are brought into a single, coherent form. Secondly, two previously distinct domains—relativity and quantum mechanics—are unified without remainder. Thirdly, this unification is achieved in a form that exhibits precisely those features—simplicity, symmetry, generality, and truth—that constitute mathematical beauty.

The aesthetic judgment here is not an ornamental addition to the physics but a recognition of the way in which a complex reality has been brought into an intelligible and ordered form. The equation does not merely describe the behavior of electrons; it does so in a way that displays the coherence of the underlying structure itself.

What Other Physicists Say About Theoretical Beauty

Many physicists and mathematical physicists involved in the investigation of the behavior of the basic constituent forces and particles of the natural world have often appealed to aesthetic principles in justifying their selection of theories. It is hard not to at least acknowledge the possibility that their deep acquaintance with the practice and outcomes of their theoretical work may not have brought them face-to-face with an empirical reality which escapes measurement precisely in virtue of its normative nature, but is, contrary to the contemporary zeitgeist, ontologically objective despite that.

Consider the following observation by Nobel laureate Subrahmanyan Chandrasekhar in his article "Beauty and the Quest for Beauty in Science." He begins his exposition by suggesting that "all of us are sensitive to nature's beauty. It is not unreasonable that the natural sciences [share] some aspects of this beauty."[75]

The Nobel prize–winning physicist Steven Weinberg, in a much-quoted sentence in his book *Dreams of a Final Theory*, says, "We are discovering laws that are becoming increasingly coherent and universal, and beginning to suspect that this is not merely an accident, that there is a beauty in these laws that mirrors something that is built into the structure of the universe at a very deep level."[76]

Of the many authors who have written full-length works extolling the virtues of beauty which they make explicit belong *to* the phenomena they investigate, a small selection might include Nobel laureate F. Wilczek, *A Beautiful Question: Finding Nature's Deep Design*; A. Zee, *Fearful Symmetry: The Search for Beauty in Modern Physics*; M. Livio, *The Accelerating Universe: Infinite Expansion, the Cosmological Constant, and the Beauty of the Cosmos*; Ian Glynn, *Elegance in Science: The Beauty of Simplicity*; and Nobel laureate Leon M. Lederman, *Symmetry and the Beautiful Universe* (with C. T. Hill).[77]

75. Chandrasekhar, "Quest for Beauty."

76. Weinberg, *Dreams of a Final Theory*, 243.

77. Wilczek, *Beautiful Question*; Zee, *Fearful Symmetry*; Livio, *Accelerating Universe*;

The editors of a special issue of *Philosophia Mathematica*, titled "Aesthetics in Mathematics," begin their introduction to the journal saying,

> Mathematicians often appreciate the beauty and elegance of particular theorems, proofs and definitions, attaching importance not only to the truth but also to the aesthetic merit of their work. . . . Others regard mathematical beauty as a key motivation driving the formulation of mathematical proofs and *even as a criteria for choosing one proof over another.*[78]

The explanatory power of mathematics in physics has added to the compulsion felt by many theoretical and mathematical physicists to ascribe beauty to equations and mathematical processes. D. J. Gross, the physics Nobel Prize winner for 2005, comments in a paper delivered to the American National Academy of Sciences in 1988 that

> for the fundamental Laws of physics, we need deep mathematics and as we probe deeper to reveal the ultimate microscopic simplicity . . . we require deeper mathematical structures. These structures are not only deep but also . . . beautiful and powerful.[79]

Nobel theoretical physicist Frank Wilczek's comment on the Dirac equation that it was "achingly beautiful"[80] appears to have been prompted by the link between its simplicity and explanatory scope and its "deep structure."

Chandrasekhar repeatedly argued in a number of works that beauty has often been associated with truth in the physical sciences, particularly where mathematical simplicity, symmetry, and explanatory power converge in successful theories.[81]

Neils Bohr, P. A. M. Dirac, Albert Einstein, W. Heisenberg, T. D. Lee, M. Gell-Mann, and H. Weyl are just a few among many other examples of notable Nobel physicists who have made similar declarations.[82]

In many places in this book, I have mentioned scientists and others who have made references to the aesthetics of Einstein's work. It is worth here reflecting on why this is so, and what Einstein himself made of the question of beauty in physics. An author with a particular interest in these

---

Glynn, *Elegance in Science*; Lederman and Hill, *Symmetry and the Beautiful Universe.*

78. Breitenbach and Rizza, "Introduction," 153; emphasis added.

79. Gross, "Physics and Mathematics," 8372.

80. Wilczek, as quoted in Farmelo, *Strangest Man*, 312.

81. See Chandrasekhar, "Quest for Beauty," and Chandrasekhar, *Truth and Beauty*, esp. ch. 3 ("The Quest for Beauty").

82. Engler, "Aesthetics in Science and Art," 24.

questions has written on the issue. In a paper titled "Einstein and the Most Beautiful Theories in Physics," Gideon Engler makes the case that, despite some theories eliciting disagreement as to their beauty,

> there are two fundamental theories for which such a difference of perception is inconceivable. These are Einstein's Special Theory of Relativity and the General Theory of Relativity. . . . Scientists unanimously praise these theories for their beauty to the highest degree.[83]

Engler has looked in detail at Einstein's use of aesthetic criteria in the special and general theories of relativity. A small literature surrounds this issue. Whatever the intricacies surrounding the discussion, the debate suggests it is at least plausible that Einstein's references were not simply emotive or metaphorical in nature but represent genuine constraints on theory. One of Einstein's biographers, Steve Gimbel, suggests that

> for Einstein, beauty was a mark of truth, and unity and coherence were a mark of beauty . . . so he worked to unify all of physics into a single comprehensive set of equations.[84]

Given the many encomiums heaped upon beauty in physics by scientists, and the seriousness with which at least two of the greatest physicists of the twentieth century took the issue, there is at least plausible evidence to support the hypothesis that mathematical physics *reveals* that beauty participates in composing the microstructures of the physical. As I have asserted, all structures have some aesthetic properties, as a matter of metaphysical necessity, but that the basic structures of the physical are beautiful rather than otherwise is a matter for empirical discovery, and this is what physics seems to support.

Let me end this section with a forthright view on the place of beauty in physics expressed by J. McDonnell that probably encompasses many of the thoughts of some of the physicists I have quoted and discussed:

> A beautiful mathematical theory will win out ultimately, though not necessarily in the short term. An ugly theory which manages to explain a seemingly unrelated phenomena and to explain the experimental data may win out in the short term, but can only ever be a stepping stone. A beautiful theory will allow us to uncover deeper phenomena. Mathematical beauty is a better guide to ultimate truth and there are occasions when we need to

83. Engler, "Einstein," 27.

84. Gimbel, *Einstein*, 152.

> keep working on a beautiful theory, even when it does not agree with experiment, in order to learn what it is trying to tell us.[85]

Can Mathematical Beauty Lead Physics Astray?

The last section having been an intoxicating foray into the beauty of the sub-particle world and its forces, it would be neglectful not to discuss a physicist who is skeptical that the aesthetic criteria contemporary physicists have lauded may, in fact, have led physics astray by distancing them from its necessary attachment to experimental evidence.

Indeed, S. Hossenfelder's book on the issue is titled *Lost in Math: How Beauty Leads Physics Astray.* She has defended related views online and elsewhere, including in an accessible essay titled "Beauty Is Truth, Truth Is Beauty, and Other Lies of Physics."[86] It is not possible here to develop a full critique of her position, which in any case demands extensive familiarity with the physics and is partly an empirical disagreement that will ultimately be resolved within physics itself. What I hope briefly to accomplish here is to place her criticisms in context and indicate why they do not undermine the broader claims advanced in this thesis.

Hossenfelder's objections are directed against what she argues has become an established tendency within theoretical physics to privilege aesthetically attractive hypotheses. She is primarily concerned with research programs involving supersymmetry, the multiverse, and grand unification, which, she maintains, have involved decades of investigation without decisive empirical confirmation or major breakthroughs in the foundations of physics. Some proposals, she points out, appear difficult, and perhaps impossible, to test experimentally because they concern realities that may in principle be unobservable.

She identifies three principal aesthetic criteria that have guided much recent theoretical work: simplicity, naturalness, and elegance. It is important to note that her objections are largely confined to these particular aesthetic ideals, which she argues have come to dominate contemporary thought in physics. One may dispute the precise definitions she offers of each, but Hossenfelder suggests that they derive from the extraordinary success of current foundational theories, such as the standard model and general relativity—theories she nevertheless acknowledges to possess striking forms of beauty. Physicists, she argues, naturally hoped that still more fundamental

85. McDonnell, *Pythagorean World*, 20.

86. Hossenfelder, *Lost in Math*. See also Hossenfelder, "Beauty Is Truth."

theories would exhibit similar aesthetic features, but the strategy has thus far yielded disappointing results.

Despite the polemical force of her titles, however, Hossenfelder concludes on a more cautious note. The laws of nature presently available to us, she argues, remain incomplete. A satisfactory understanding of the quantum behavior of space and time may require profound revision either of gravity, quantum physics, or both. Yet she nevertheless expresses confidence that another major breakthrough in physics will eventually occur and that it, too, will possess a kind of beauty.

Hossenfelder may therefore have a point. Future theories may not exhibit beauty in precisely the same way as general relativity or the Dirac equation. The expectation that nature must indefinitely conform to a particular conception of elegance, simplicity, or symmetry may indeed prove misleading. Yet this possibility is not a difficulty for my thesis. My argument has never been that specific aesthetic preferences provide infallible guides to discovery nor that mathematical beauty functions merely as a methodological rule imposed upon physics from outside.

As McDowell argues, the world is not encountered as a bare, unconceptualized reality upon which human meanings are subsequently projected. It comes to us already within the space of intelligibility. My proposal has been that mathematical physics represents one of the most profound forms in which this intelligibility becomes articulated. The aesthetic character of mathematical structures is therefore not simply an optional psychological preference of physicists, but part of the way in which structured order becomes available to rational apprehension.

This does not entail that reality must conform to one historically conditioned aesthetic grammar. The history of music provides an instructive analogy. Certain periods in music elevated particular forms of harmony, balance, and resolution into standards of beauty. Later developments often violated these expectations while nevertheless disclosing new forms of intelligible order. The movement from classical tonality to the harmonic innovations of Wagner, the rhythmic dislocations of Stravinsky, or the structures of modern jazz did not abolish musical intelligibility because older aesthetic expectations were disrupted. Rather, new forms of order emerged whose intelligibility required fresh modes of apprehension.

Something similar may prove true in physics. The failure of certain aesthetically motivated theories would not demonstrate that beauty is irrelevant to physics but only that our current expectations concerning the forms beauty must take are limited. Nature may instantiate kinds of mathematical order whose aesthetic character differs profoundly from those presently most admired. What remains significant is that the structures disclosed by physics continue to present themselves as intelligible structures capable of rational and mathematical articulation.

Mathematics in this sense should not be understood as a repository of independently existing Platonic entities awaiting physical instantiation. Rather, it is the articulation of possible structural relations through which the intelligibility of the world becomes expressible. Physics determines empirically which of these articulated structures are realized in nature. Whether the resulting structures are simple or complex, elegant or strange, symmetrical or asymmetric, they nevertheless remain open to aesthetic description precisely because structure as such possesses aesthetic character.

As physics advances, the depth and range of the mathematical structures disclosed in conjunction with experimental science may reveal forms of order very different from those currently familiar to us. Such developments would not undermine the relation between mathematics, structure, and aesthetics defended here, but enlarge our understanding of the ways in which the intelligibility of beauty may be manifested within the physical world.

In sum, Hossenfelder's critique is well accommodated within my view. Whether she is correct about particular developments in contemporary physics is a matter for the future course of physics itself rather than a decisive objection to the broader philosophical claims advanced in this thesis.

Causality and the Janus Face of Aesthetic Structures

If the contentions advanced in this chapter are correct, then aesthetic properties can no longer be regarded as additions imposed upon an otherwise value-free physical reality. The structures disclosed by mathematics and mathematical physics are not merely causally efficacious; they are also susceptible to aesthetic apprehension. We encounter them through different but related modes of disclosure: through the empirical and mathematical practices of physics, and through the recognition of qualities such as elegance, simplicity, coherence, and beauty. The world, at the level of its most fundamental structures and forces, could have exhibited forms experienced as confused, inelegant, or opaque. The remarkable fact is that it frequently does not.

That such judgments can intelligibly be made concerning the deep structures of nature is itself philosophically significant. Aesthetic judgments are normative judgments. To claim that something is beautiful, elegant, or harmonious is simultaneously to imply that it could have been otherwise: less elegant, less coherent, less beautiful. Such judgments therefore presuppose standards which transcend the merely brute occurrence of physical events. If one suggests that a mathematical structure or physical theory "could have been more beautiful," one necessarily invokes some conception of better or worse, of success or failure, against which the judgment is made.

Yet within a conception of nature understood solely as the aggregate outcome of contingent physical processes, such standards are difficult to locate. The events of the physical world simply occur as they do. They possess no intrinsic orientation toward value or meaning. And yet the structures uncovered by the natural sciences appear repeatedly to invite precisely these forms of normative and aesthetic assessment. The question therefore naturally arises as to whether beauty, intelligibility, and normativity are merely subjective responses projected onto an otherwise indifferent reality, or whether they reflect something more fundamental concerning the nature of the world itself.

This question becomes still more pressing when one recognizes that the world contains not only causal processes and mathematically describable structures, but rational beings capable of understanding and intentionally manipulating those structures. Human beings do not merely observe the contingent operations of nature. We employ them purposively. We reshape the materials and forces of the world in accordance with intention, imagination, and standards of value. In doing so we produce artifacts: objects whose structures are not simply the result of brute causal interaction but of purposive rational activity. Moreover, many such artifacts are intentionally beautiful.

The existence of artifacts introduces into the world a distinctive form of structure in which causality, intelligibility, purpose, and aesthetic value are united in a directly observable way. A painting, a cathedral, a mathematical proof, or a symphony are not merely physical arrangements of matter or sound. They are structured expressions of intentional life. How such purposive and normative activity is possible within nature, and what it might suggest concerning the ultimate character and origins of reality, forms the starting point of the next chapter.

Chapter 3

Human Artifacts and Teleological Structure

SUMMARY

THE "PHYSICAL" WORLD IS not the only dimension of nature in which structures, objects, and processes appear. We also construct artifacts from what nature makes available to us as creatures *of* nature ourselves.

What "nature" makes available to us is physical regularities and causes which we are able to manipulate for our own purposes. Those purposes are the "final causes" that explain why artifacts emerge. That nature allows for teleological interference in causal regularities cannot be accommodated within the assumptions of ontological naturalism. Teleology, one might say, is present in nature as a consequence.

OVERVIEW OF THE CHAPTER

Contrary to received wisdom, the teleological provenance of artifacts does not place them outside of nature because we are part of nature. As Lynne Rudder Baker has pithily described the consequences for the ontology of artifacts,

> The existence of artifacts depends not on nature-as-if-we-did-not-exist. . . . Since nature *has* us in it, this distinction (between nature-as-if-we-did-not-exist and nature-with-us-in-it) is

> no satisfactory basis for a verdict of ontological inferiority of artifacts.[1]

I suggest that being structures, artifacts share a fundamental ontological principle with any other object that guarantees their equivalent status: they are subject to the same a priori mathematical constraints of any structure, and like any structure they display express aesthetic and normative properties.

Our teleological creations are as much restrained by these features as are the objects that are the outcome of "physical" regularities alone. The features any physical process or item must instantiate are precisely those that our artifacts, in order to be such things, must also instantiate.

While biological processes are necessary to the *ability* of conscious rational creatures to engage with normative concepts like those of aesthetics, for example, or value, and act in accordance with them, that does not *explain* the ontological provenance of teleological and intentional standards in nature. I discuss why this is so via a critique of the work of Terence Deacon, who has attempted to expand the theoretical resources of biology to incorporate teleology.

1. Baker, "Shrinking Difference," 6.

Human Artifact 1. The Strasbourg Cathedral Rose Window, second period of construction, thirteenth to fourteenth century[2]

2. Stained glass window of the Strasbourg Cathedral (France) by W. Bulach, https://commons.wikimedia.org/wiki/File:00_1812_Stained_glass_windows_-_Notre-Dame_de_Strasbourg.jpg, CC-BY-SA 4.0, https://creativecommons.org/licenses/by-sa/4.0.

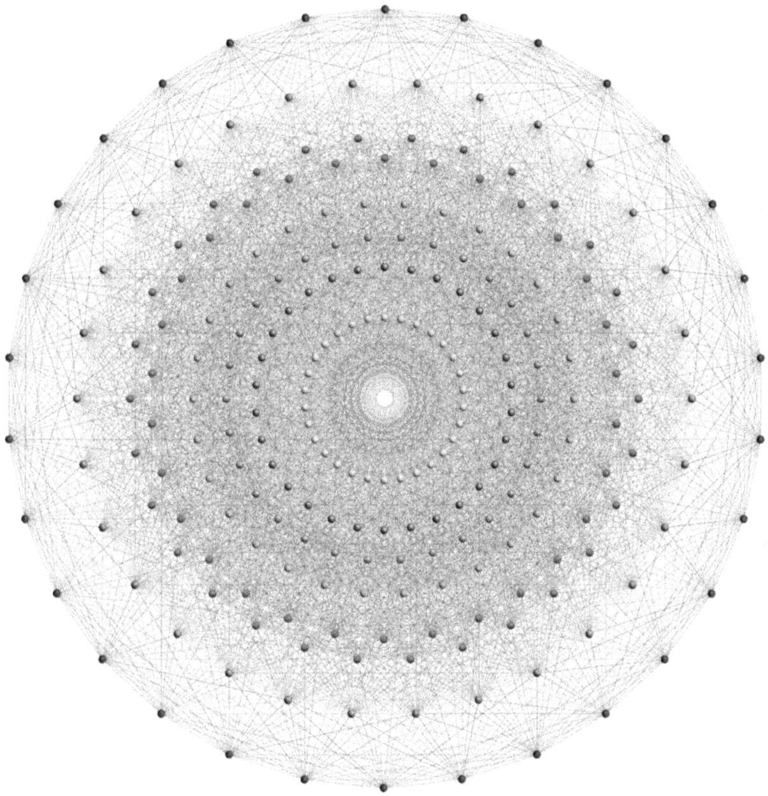

Human Artifact 2. A visual representation of the mathematical E8 Lie group, a perfectly symmetrical 248-dimensional object posited to explain many of the interactional relations between particles in mathematical physics[3]

3. Visual representation of the E8 group by Jgmoxness, https://commons.wikimedia.org/wiki/File:E8Petrie.svg, CC-BY-SA 3.0, https://creativecommons.org/licenses/by-sa/3.0/deed.en.

PART 1

What Are Human Artifacts?

The previous chapter dealt with the aesthetics of the events and items of what was called the "base" physical structure of the world: the province of particle physics. That chapter began with a quote from Wendy Pullan, the editor of a series of papers presented in book form titled *Structure in Science and Art*. Her introduction to the book begins by reminding the reader that

> bridges and buildings, DNA and the periodic table, flora and fauna, machines and circuits, human beings and the societies of which they are part, even thoughts and ideas—all of these can be understood as particular structures that are part of the world and our experience of it.[4]

It is already apparent in this listing of "structures" that what is to count as *any* kind of object, whether the outcome of human intention or the strictly causal processes of the physical world, share structure as a common and necessary condition of existence. I will refer to any structure that involves the purposes of rational reflective creatures such as ourselves as an "artifact."

The term "artifact," or what is normally taken to make up the set of such things, is often used unintentionally to refer only to "external" objects of human origin like knives, forks, vehicles, dwellings, pencils, and laptops. It should not be forgotten, however, that such things as the physical and phonetic elements that are the vehicles of human writing and speech are also "artifacts": bits of isolated physical "stuff" used as the vehicles of speech, writing, music, logic, mathematics, and other forms of meaning, symbolization, reference, and communication. Without the physical items used for these purposes, there could not be identifiably human and personal lives, let alone the pursuit of empirical or any other form of knowledge.

Scientific theories, to offer a further example of the point, rely on instruments that measure and investigate the properties, processes, and events of the natural world. Even the humble tape measure or abacus are artifacts, as well as those instruments that can extend human faculties far beyond ordinary perception, such as the capacity of a telescope to detect the light from a far-off galaxy, or that of a receiver to pick up the residual background radiation from the "Big Bang" which is fundamental to that theory. All these artifactual measuring devices have been intentionally conformed to enhance what we cannot do with our own biological perceptual resources.

4. Pullan, "Structuring Structure," 1.

It is a matter of such ingrained familiarity in determining a human life to be what it is, that it is sometimes hard to recognize how remarkable it should be that *Homo sapiens* using rational and imaginative faculties can make the objects, substances, and processes of the external "objective" world into *entirely* novel things that perform certain functions which are nowhere to be found in any listing of the items of any natural science.

Distinguishing Artifacts from the Objects of Science

What distinguishes artifacts from the objects of "natural" science? "Natural" objects, the philosopher Lynne Rudder Baker conventionally observes,[5] come into being without any human intervention, while artifacts depend *ontologically, not just causally,*[6] on human purpose. They are "intention dependent,"[7] as she calls them, and consequentially could not exist in a world without minds possessed of beliefs, desires, and intentions that bestow their functions on them.

The intention-dependence of artifacts introduces the idea of "final cause" into the causal explanation of their origins. Any human creator must have a preconceived item in mind (which has no existence in the "natural" world) which they wish to bring to existential fruition. That item might be characterized as a final cause: it provides the intended outcome in virtue of which the materials that come to constitute the artifact are organized to bring it about. To do something in order to achieve a "purpose" is a fundamental principle of teleological explanation.

Any artifact of any kind also requires some prior understanding or skilled familiarity, whether of the practical kind or the theoretical kind, of the properties and possibilities of the natural materials with which its construction is associated: a rational familiarity, in other words, with the structure and causal behavior of the natural elements of which any artifact is to be composed ("If I do *x* with physical substance §, *y* will happen. I do not want *y* to happen so I will not do *x*"). It is, in other words, the bringing together of "physical" elements to fulfill a conceived purpose.

5. Baker, "Shrinking Difference."
6. Baker, "Shrinking Difference," 3.
7. Baker, "Shrinking Difference," 3.

The Degradation of Artifacts in Ontological Naturalism

Objects of human creation, Rudder Baker points out, have often been degraded in metaphysical and ontological thought by being denied the cherished ontological status of "substances." Without entering into the depths of this discussion, the intuitive thought behind this is that because they are the products of mind and intentional agency, they somehow lack the full "reality" of natural objects.[8]

Their "artificiality" regularly excludes them from any ontology of what "really" exists. Without human beings around to use them, it might seem apparent that a "fork," a "brake," and a "painting" are of no significance when it comes to listing the components of the ultimate existence and functions of physical objects and events: when the fundamental constituents of reality are described by the natural sciences, human artifacts will be excluded because of their teleological origins.

This dismissal is of a piece with the thought that any aesthetic characteristics they may have will be dependent on how they are perceived by their human creators, not on the intrinsic characteristics of their structures.

Ontological Naturalism and Teleological Artifacts

These outcomes are consistent with the picture of the world as ultimately "physical" and "impersonal" which I have been concerned to contest: if "materialism"/"physicalism" describes the primary or "real" substances out of which everything is constituted, and to which all other entities and processes must be reduced or otherwise made dependent, then artifacts can have no proper ontological status when we come to list the components of what ultimately "exists," nor can they have an intended beauty; they are ultimately dependent for their shapes, functions, and physical structures on the *projective* agency of human thought and life.

Because the division of world into "personal" and "impersonal" maps onto the difference between nature as "physical" and "personal" as teleological, expressive, and normative, it is relatively straightforward from the physicalist perspective to exclude artifacts as bearing in their "natural" state marks of intention, meaning, and purpose: a "painting" or a "symphony," a

8. The extrusion also has the rarely mentioned consequence that it is difficult to show how science itself can represent the world *and* be an item of the world itself. If a theory is an epistemological artifact that falls entirely into the personal part of the world, then what the threads are that tie the theory to the world are a mystery.

"fork" or "bed," or even a "theory," is just, in the end, a collection of natural entities aggregated together in such a way that we call them an *x*.

That aggregation, however, will not delimit anything of real ontological significance other than our gathering similar aggregates under a collective name to serve our purposes. "Meaning," "purpose," and all the cousins of teleological life on display in the various artifacts of personal life are projections from the subjective interior of human consciousness onto brute physicality: a beautiful painting which displays a delicate scene of human domestic interaction, for example, that captures something of the profundity of human existence in the very ordinary, is nowhere to be found in its physical construction.

Should we accept the distinction between the personal and the impersonal in ontological naturalist terms, or is its apparent incontestability not an outcome of scientific practice but metaphysical prejudice? I have argued that it is indeed the latter: the reunification of "purpose" and the "natural," I suggested, requires an overarching metaphysics that can incorporate both into a unified picture of the world. Part of that exercise is to show how the "artifacts" that participate in making human life "personal" and teleologically comprehensible have an ontological significance equivalent to the "physical" events and properties of the "natural" sciences.

"Real" Objects: "Nature-with-Us-in-It" vs. "Nature-as-If-We-Did-Not-Exist"

Lynne Rudder Baker offers us a reflection that can move the argument in the required direction. The distinction between artifact as "not-real" and natural item as "real" rests, as Baker points out, on the tendentious claim that what makes a "real" object a substance is that the condition of membership is "set" by nature and not by us.

> Of course, the existence of artifacts depends on us: but we are part of nature. . . . The existence of artefacts depends not on nature-as-if-we-did-not-exist. . . . Since nature *has* us in it, this distinction (between nature-as-if-we-did-not-exist and nature-with-us-in-it) is no satisfactory basis for a verdict of ontological inferiority of artifacts.[9]

The restriction of artifacts to human agency is in any case more of a *distinction* between different teleological capacities that various living things can have in the world, rather than a *dichotomous* division between artifacts

9. Baker, "Shrinking Difference," 6.

and natural objects, than the case for "material" or "physical" ontological supremacy cares to concede.

As any brief reflection on the animal world will show, all living creatures exploit their environments to create "artificial" structures like nests, holes, dams, webs, and, in the case of chimpanzees, elementary tools, like sticks inserted into ant heaps to extract the insect. In this very fundamental sense, they are also engaged in basic teleological activity. No one would question, however, their full status as proper occupants of a scientifically accessible nature.

With the evolution of human rationality, *Homo sapiens* takes a step up the ladder of teleology. He is able to create novel items that emerge from his rational, inventive, and imaginative capacities and are not just thrust upon him by his animal nature,[10] as in the case of the bee's nest or the beaver's dam.

Homo sapiens becomes responsive to the normative properties available to him in the world to rationally design structures of functional use and even beauty. These extend from using physical items to symbolize any number of things and relations in language and other forms of symbolization to independent artifacts. Human beings can now even construct artificial elements that provide additions to the world of natural elements.

A nice example of this comes from an edition of the *New Scientist*, which I quote in full:

> Diamond, one of the hardest known materials, could have a rival. A new theoretical form of carbon called pentadiamond is just as hard as the gemstone, but as light as pencil graphite, and could theoretically be used for storing gas.[11]

In this case we have even added to the natural world a new type of molecular structure.

Rudder Baker's analysis of "nature-with-us-in-it" has two fundamental *logical* consequents: Firstly, if we are taken to be a part of "nature," then it follows that nature must *include* those features which are commonly taken to distinguish us from the purely physical. Secondly, if the products of our agency reflect rational teleology in their identity and any account of their functioning ("the purpose of a chair is to sit on"), it follows that nature must

10. The position is somewhat more nuanced than this, however. Some animal behaviors have appeared only in particular local contexts. Some octopuses carry around with them halves of coconut shells discarded by local fishermen in a couple of their eight arms and can assemble them into shelters if threatened by predators. I owe this point to Dr. D. de Pomerai.

11. *New Scientist*, "News in Brief," 18.

have a proper place in any account of its basic ontology for teleology and what is required for it.

Of course, one might deny that human rationality and teleological agency are themselves anything other than illusions of the neural activities of the frontal cortex. Or one might accept that teleological language has some functional utility in human behavior. Nevertheless, it might be argued that despite teleological concepts serving useful functions for us, they are ultimately replaceable by, or reducible to, supervenient on, or, in some way to be fully determined, dependent on the purely causal language of some science or sciences.

If one does deny teleological reality, however, the burden must fall on the protagonist to explain why the denial is not itself a manifestation of precisely what it is concerned to deny: a teleological and meaningful argument expressed in physical and abstract vehicles (syntax and symbol) that are purposefully organized to convey meaning. In other words, to assert the case, the protagonist of the view must precisely exploit those properties which are denied, to speak intelligibly at all, let alone formulate the hypothesis that teleology reduces to the causal regularity of physical items.

The Incompatibility of "Nature-with-Us-in-It" and Ontological Naturalism

The distinction that Baker illuminates between nature-as-if-we-did-not-exist and nature-with-us-in-it is taken by the ontological naturalist to discriminate between what has ontological priority and what has not. The dichotomy he evolves from it is a metaphysical judgment emerging from a productive principle of methodology, not an empirical fact.

Here is the metaphysical judgment to which he thinks the method commits him: The totality of anything with a physically characterizable nature—the particles and forces of physics, the structures and properties of molecules and chemical elements, the formation of rocks, seas, stars, and mountains and organic life—are all dependent on the regularity of purely causal physical events. The behavior and nature of all these constituents are fully described by "laws of nature." The laws of nature present a "closed" world completely explained by their efficient causal and physical relations.

Or, as I quoted Robert Hanna as describing ontological naturalism, "the *entire* world including living organisms, animals, human beings, and persons, operates according to non-teleological, mathematic-physical principles alone."[12]

12. Hanna, *Kant, Science, and Human Nature*, 8.

This metaphysical conclusion leaves unanswered the question of how the objects of human creation (or those of other creatures) are to be ontologically characterized, other than an assertion (which I have strenuously denied) that mathematic-physical principles are obviously nonteleological. Teleology is, by implication from this principle, not methodologically and metaphysically respectable.

If we do want to take the intentional causation of human artifacts seriously, then there is a real question about their identity and place in the ontology of the world which derives from two considerations which are apparently at explanatory odds with each other: Firstly, written shapes, objects like tables, and functioning devices like computers and trains are fully physical items, properly part of anything that might be called "the physical world." In this respect, the parts that make up any artifact are subject to the disciplines which investigate whatever physically composes them.

Secondly, on the other hand, there is no science peculiar to knives and forks, chairs and tables, heaters, thermometers, trains, the visual representations of alphabetic structures and elements, institutions, planes and carriages, and hadron colliders; these are not the subject of any dedicated scientific discipline insofar as their identity or functions go.

These two facets of humanly derived artifacts can create a difficulty for how they are conceived if the naturalist principle outlined above is applied to their identity. We can approach the problem by recalling that the extrusion of "final causes" and "purposes" from explanations of natural events is part of the origin of productive scientific method. Human artifacts, however, are structures whose nature and functions are *determined* by the purposes for which they are made: their ultimate origin has an intentional cause, but their operations are entirely determined by the "closed" causality of the physical world.

Teleology and Ontological Naturalism: Intention and Causation

The trouble emerges with the following thought: If everything is ontologically finally determined by the regularity of causal events as described by the laws of nature, how is it possible that human artifacts can be intentionally put together to create things which have not emerged by the operation of those laws alone? How does the "physical" world submit to purposeful interference by ourselves to create that to which it has not given rise by "natural" processes?

Or to put it another way, how are we able to devise and use things for our own purposes which redirect causal laws to change physical items into artifacts that fulfill those purposes? If nature is fully determined by efficient causes, where is the ontological space for the construction of artifacts within it which are the outcome of prior intent? How can an *intention*, a *nonphysical final cause*, a property of mind, contribute to bits and pieces of nature becoming a "fork"? There are no "laws of nature" that have forks among their theoretical components.

The laws of nature are typically taken to describe purely "objective" empirico-physical and regularly recurring events of the world conceived under one aspect of its presentation to us. It should be apparent that laws of nature thus conceived are not sufficient to conceptualize all the properties of a natural world, what it can give rise to, and what place the laws have in relation to rational agents like ourselves who belong in it but are able to transcend its physical nature to create novel objects.

And yet it seems to be an undeniable a posteriori truth about our experience of the world that it has the curious property of being exploitable by reason and purpose that cannot be accommodated within the concept of the "laws of nature." I say "a posteriori" because it is a truth about our experience of the physical world that this is possible, but not one garnered from the outcomes of scientific activity.

How *could* it be since the very possibility of science requires physical vehicles to communicate scientific truths like mathematical symbols, logical relations, and scientific instruments for its purposes, all of which are artifacts and all of which have properties, purposes, and functions which transcend the regularity of physical causation.

The possibility of science, in other words, is dependent on it being the case that the physical world can be shaped by our purposes into what is necessary *for* science. Laws of nature explain and predict the operations of entities and change conceived causally and measurably, not the production of items which are intentionally produced, particularly those scientific instruments that measure and describe those very operations.

An ability to exploit the laws of nature cannot be itself a law of nature since if it was a law of nature, the idea that a "law of nature is being exploited" would have no meaning. To exploit a law requires a self-conscious recognition of that law and what it entails, then rationally putting it to use. That we are able to unearth, understand, and use laws to change the physical world about us to accord with purposes only in our minds can be understood if we are willing to accept the possibility that underpinning all of the natural world is a structure that embodies teleology and intention and displays properties amenable to teleological manipulation. It is from this vantage point that we can manipulate causal regularities to our own advantage (or disadvantage).

Teleology and Change at the Structural Basis of *All* Events

That nature can be manipulated teleologically is surprising. It is only surprising, however, if one begins from the premise that teleological principles can have no fundamental role to play in the ontology of natural objects. But if one to the contrary supposes that all-natural processes and objects have metaphysical principles, as a prerequisite of their being what they are, matters are different.

When objects and events are brought into existence, they must fulfill certain metaphysical conditions of instantiation. I noted in an earlier chapter the concept of "structure" is one of those fundamental to a world that can be experienced, is contingent, and is investigatable. It is a further prerequisite to any physical world that structures that can and do change. Indeed "change" is a concept without which the critical concept of efficient "causation" has no purchase: the very idea of a dynamic "natural structure" open to scientific investigation is that it instantiates regular causal changes.

There are also certain ways in which the relationships between the parts and wholes of a structure are necessary for causal processes: causal events typically are a kind of relation between the elements of structure, those that constitute it—as, for example, the field and particle relationships whose causal interactions are necessary to articulating atomic structure—or the causal relations between properties of *different* kinds of structure. It is these relations that are answerable to, and described by, mathematics.

Of course, human creations are subject to the same conditions: no human artifact has immunity from change nor could the whole structure come into fulfilled being before the existence of its parts. Again, no human structure could fail to obey some law(s) of mathematics. Creations and artifacts are, as the objects of physical nature are, dependent on the a priori principle that the structures of nature can be changed.

The capacity nature has to be changed both by regular causal processes and teleological manipulation constrains any possible items of existence in the same way: any functions, shapes, dimensions, aesthetics, and causal possibilities have to obey some law of mathematics to form a logically possible structure, whether teleological in origin or the outcome of purely "physical" causes. These modal suggestions are of a piece with the assumption that any world must instantiate certain dictates of metaphysical possibility to have any existing physical forms, whatever these might finally be agreed to be.

Nothing in those constraints, however, inhibits the fact that we *have* constructive powers of corporeal (and abstract) teleology. What it does inhibit is what can be done *with* our powers. One might want to think of this as an a posteriori truth: since we do construct "artificial" objects out of physical materials, the materials in question must have the property of being able to be manipulated. Most critically, if it were a metaphysical truth that nature was "physically closed" by the regularity of its causal processes, it is prima facie hard to conceive how human rationality and corporeal agency could intrude into that closed circle to create designed objects.

PART 2

The Theoretical Role of Teleology in Biology

The natural world allows for "natural" elements to be composed into new shapes and forms by rational creatures of its own production: *Homo sapiens*. As author and designer, we are provided for by nature for the task of teleological construction, able to deploy thought and reason in pursuit of our goals, and physically and cognitively equipped to manipulate what lies naturally before us. Our rational engagement with the world is also of course via our biological constitution.

As one might expect as a result, the discipline in which the idea of teleology has proved most obviously troublesome, then, is that of biology. The *Stanford Encyclopedia of Philosophy* begins its optimistic contribution to the topic in the following way:

> The manifest appearance of function and purpose in living systems is responsible for the prevalence of apparently teleological explanations of organism structure and behavior in biology. Although the attribution of function and purpose to living systems is an ancient practice, teleological notions are largely considered ineliminable from modern biological sciences, such as evolutionary biology, genetics, medicine, ethology, and psychiatry, because they play an important explanatory role.[13]

The description offered by the encyclopedia is sanguine about the explanatory role of teleology, but how "teleology" is to be understood in scientific terms is still a contentious issue. Under the rubric "teleonomy," a distinct literature has built up offering naturalist attempts either to accommodate "teleology" within a naturalist worldview or find ways of preserving its explanatory functions but denuding it of its teleological language.[14]

An abiding concern of many contemporary theorists in the cognitive neurosciences, cognitive anthropology, and biology is the worry that

13. Allen and Neal, "Teleological Notions in Biology."

14. Allen and Neal, "Teleological Notions in Biology."

teleology may let Cartesian consciousness back into their disciplines. Since all neurobiological and cognitive neuroscientists claim to be no longer influenced by Cartesian metaphysics, the anxiety they have is comprehensible. The emphasis placed on this denial is to stress their formal identification with the naturalism which is what they consider follows from the methodological practices of the sciences. I critiqued this metaphysical conclusion in an earlier chapter.

The rejection of teleology in experimental science in the modern era evolved with the growth of the "mechanical" philosophy of the Cartesians and the Epicureans in the seventeenth century (see the discussion below). The rejection of teleology entails, of course, the rejection of "design." This latter concept is inseparable from the notion of a constructing and reflective mind that builds within itself an imaginative picture of the item that the person is concerned to bring into reality. The "designed" picture may be thought of in Aristotelian terms as the "final cause" or purpose that governs the eventual emergence of the designed object. "Final causes" were rejected by Descartes and many others in the genesis of the modern sciences, as I discuss shortly.

Notably, however, one of the leading scientists of his day, and a classic figure in the development of the discovery of the elemental nature of things, Robert Boyle (1627–1691), offered a defense of the usefulness of teleological inference.[15] The "father of chemistry" argued in his monograph for a place for final causes, in *A Disquisition About the Final Causes of Natural Things Wherein It Is Inquir'd, Whether, and (If at All) with What Cautions, a Naturalist Should Admit Them?*[16]

Boyle, within his own emergent discipline, recognized that it is indeed hard to discuss nature without teleology intruding at some point. The lesson to be drawn is that such concepts entailing design in nature did not appear to at least one of the early progenitors of scientific method, to be obviously discredited by the new developing experimental and theoretical methods.

Descartes, "Mechanical" Structure, and Teleology in Biology

The denial of Cartesian dualism insists that the division between matter/material and consciousness/mind is primary. The other equally important Cartesian distinction is that between the mind as rational and teleological *and* the "external" world it engages with. The latter in opposition to the

15. For a discussion of this and Boyle's work on final causes, see Lennox, "Robert Boyle's Defense."

16. Boyle, *Disquisition About the Final Causes.*

former is conceived by Descartes to operate like a "mechanistic" device of his time—a clock, for example. This division is one which precedes the Cartesian dualism of mind and body and is arguably of equal metaphysical significance.[17]

Nevertheless, the intuitive reality of the role of reason and teleology in understanding behavior in biology through terms like "function," "in order to," "for the purpose of," and "for the sake of" often appear to be the only available concepts to describe certain phenomena. "Maturation," for example, is a fundamental concept without which elements of a developing lifespan and its goals cannot be described.

An array of cognitive and behavioral structures—such as those to do with working, short-term, and long-term memory, "social representation," and so on—all require reference to the purposes they fulfill in the life of a creature on pain of that life being unintelligible. Any form of memory that influences a future action, for example, is entirely cognitive, representing a past reality that no longer has any "physical" existence but can provide a purpose that causes future behavior. Similarly, social representations provide goals that can determine even instinctual behavior. Bees, for example, perform complex social movements that indicate the position of flowers and their distance to other hive bees.

The need for explanations of action based on an organism's capacity to represent a future outcome to itself which it acts to bring about, puts pressure on a fundamental principle of physicalist naturalism. As I outlined earlier, for the physicalist naturalist it is an ontological necessity that nothing in the physical world should escape the regularity of efficient causation, *x* preceding *y* as a matter of uniform occurrence.

A few biologists, neurobiologists, and philosophers of biology have labored to gain some clarity on the acceptability or otherwise of teleological language in sentient creatures as a whole. A contemporary example is to be found in the extensive work of Terrence Deacon, a neuroscientist and biological anthropologist. He explicitly holds Descartes to account for his conception of the external physical world as "mechanical." The mechanical conception, he argues, displaces the need for a necessary space for teleological explanation in the behavior of sentient organisms, among whom are included ourselves.

Deacon understands the need for "science" to respond to the question of teleology and purpose-directed causation at least in the biology of sentient organisms. I question, however, whether his restriction of teleological ontogenesis to biological organization, whether in the form of

17. Descartes, *Meditations*. See also Dicker, *Descartes*, and Williams, *Descartes*.

neurological tissue or otherwise, is adequate to tackle the Cartesian "mechanism" problem.

Origins of the "Mechanical" World: Descartes and Aristotle

In order to understand the basis of Deacon's critique, it is useful to take a step back and examine the origins of Descartes's conception of a "mechanical" world. This will also serve the useful function of beginning to demonstrate how that picture, even with its modern updates, does not escape the basic Cartesian schema. Because of this, modern ontological naturalism is unable to articulate a proper metaphysical place for normative and expressive properties which have conceptual connections to teleology.

In her book-length study of the contribution of Descartes (1596–1650) to the dismantling of the Aristotelian theory of substantial forms and final causes to the development of a modern empirical science, Helen Hattab writes,

> [Descartes] justifies the replacement of substantial forms . . . by denying the distinction between natural and mechanical phenomena. Automata, he insists, are also works of nature. When building them we simply apply active things to passive things as we do when we sow grain or breed a mule. The difference between machines and natural objects is simply one of degree for Descartes.[18]

As she says, Descartes's view,

> in other words, [was that] since nature is just a more complex machine the same procedures that apply in mechanical demonstrations apply to natural phenomena.[19]

In the historical context Descartes was engaged in a theoretical struggle with the defenders of the Aristotelian or neo-Aristotelian explanations of natural phenomena.[20] Aristotle, or his followers, made a discrimination between "mechanics," the construction of human devices that involved the use of Euclidean geometry, and the theoretical four-cause explanation of all natural "forms" and "substances," the inanimate and animate included.

18. Hattab, *Descartes on Forms*, 85.

19. Hattab, *Descartes on Forms*, 86.

20. Aristotle's *Quaestiones Mechanicae* provided a series of reflections on a variety of simple mechanical devices which are said to have their ultimate cause as the properties of a circle. See Hattab, *Descartes on Forms*, and also Gerber, "Descartes, Mechanics and Mechanical Philosophy."

The work purporting to deal with Aristotle's discussion of elements of human mechanisms became available in Latin translation in 1519 as *Quaestiones Mechanicae*. In this work there are a number of comments about human constructions, but these seem not to have been brought by Aristotle within the scope of his "essence," "substance," and "four-cause" theory.

It is a question of historical interest why the Aristotelian discussion of mechanism did not incorporate his four-cause and substance theory. That question cannot be addressed here, but one possibility is that Aristotle's demarcation hinged on the idea that human artifacts could not appeal to the "natural" unity or "essence" of a substance occurring "naturally." For Aristotle, inside—to speak metaphorically—of every natural thing a principle, a final cause, determines how that thing is to unfold and develop within the world. Artifacts do not have final causes in the same way that govern the inner development of their structures to become the kinds of things they ought to become—a bee, an oak tree, or indeed a human being. In the context of the understandings of the natural world and its operations at the time, this is clearly an obvious option, and one which still serves as a contemporary distinction of use.

Hattab suggests that while opposing the substance theory of Aristotle, Descartes may nevertheless have been influenced at some point by Aristotle's discussion in the *Mechanicae* of the use of geometry in the construction of artifacts. Descartes seemed to have discarded Aristotelian forms and essences because they could not be measured or otherwise given useful empirical form but recognized that Aristotelian mechanics and its discussion of the uses of geometry in constructing artifacts could be usefully applied to understanding how the natural world might work.

His fundamental contribution was the extension of the idea of "mechanization" to physical objects and events in the external world as a whole. The essence of his view was that by drawing a similarity between machine and nature, it could be concluded that the material world evinces only those properties immediately susceptible to the mathematics and measurement of his time: the dimensions describable by Euclidean geometry, length, weight, and so on.

The universe and its occupants, he opines, is like a gigantic machine with interlocking physical shapes that transmitted causal relations via their immediate physical proximity. His picture of the world was that of a "plenum": a space fully filled with objects decomposable into those shapes allowed for by Euclidean geometry—causal arrangements resulting from transmission of movement accomplished by the forces exerted when one shape came into contact with another.

"Matter" in Descartes's conception and those of the new developing view of matter treated it as "lifeless and inert."[21] Recognizing, however, that a place needed to be found for human rationality and consciousness, Descartes proposed a second fundamental but different substance in interaction with the corporeal to give a unified account of the conscious human person. The external physical world was thus composed, in Descartes's view, of an ontologically different "stuff" metaphysically and scientifically incapable of supporting consciousness and rationality.

Descartes purged the world of normativity and teleology to clear it for "objective" properties (size, shape, dimension, and so on; those being the properties with which the mathematics of the time could cope). Needing a place for consciousness and rationality, he confined all of these phenomena to the "substance" of mind. The machinelike view of the world then developed through the hands of philosophers like John Locke (1632–1704),[22] eventually mutating into ontological naturalism with the place of consciousness, teleology, and value in broader reality a metaphysical mystery.

In his haste to remove the obstructions of Aristotelian theory from the "natural sciences," however, Descartes ironically removed the very features of normativity from the physical world required for the mathematical description of the causal relations of structures needed for a mathematical science. I explore and develop this idea in the next section.

Descartes's Error: The Mathematical Structure and Normativity of "Nature"

To argue that anything which is an ontological target of empirical investigation must necessarily have certain kind of metaphysical properties, is not a claim about the contingent character of a metaphysically necessary property in any empirical case: *everything* that is the subject of a scientific theory must have a structure and its associated properties.

21. Ashworth, "Christianity and the Mechanistic Universe," 61. Ashworth offers an account of Descartes's mechanical philosophy and its place in the development of the "new philosophy of nature," incorporating discussions of figures such as Pierre Gassendi, Descartes's philosophical interlocutor, and Robert Boyle, the father of chemistry. See also Broman, "Matter, Force," in the same volume.

22. See Locke, *Essay Concerning Human Understanding*. Locke was one of the principal figures in the development of empiricism, the view that knowledge arises through sensory contact with the world. He was also among those who introduced a distinction between experiences thought to mirror the world and those properties regarded as constructions of the mind.

What those structures are and how they work is the business of science. We are blind to photosynthesis, for example, before we understand the structures of solar and chemical energy transformations and the role of biological elements in that process, but that they *are* changeable structures is a condition of their appearance as physically investigatable phenomena.

To the extent that mathematics imposes upon worldly structures logical dimensions that they must obey to qualify as a physical item at all,[23] the "physical" and its fundamental base already display properties that are nonphysical, and normative. Contra Descartes, to conceive of anything *as* a "mechanism" in the terms that he does is to conceive of something which has the empirically investigatable nature that it does because it *meets* the prior conditions of manifesting as worldly things of a particular structural type.

Possibly because these metaphysical requirements of the existence of a physically functioning world share with Aristotle a demand for a priori preconditions of identification and existence, Descartes failed to make the necessary distinction as indeed did Aristotle, conflating conditions of what something has to have to have *any* physical form at all with the contingent properties that anything *has* as an extant physical structure.

Although the word "structure" is not a metonym for an Aristotelian form, understood in a similar way, it does a metaphysical job that aligns natural and human artifacts. That the world has structures with potentials to become other things, change other things, and be changed by other things is a metaphysical sine qua non of the broader picture: they are part of the very meaning of what we call the "contingency" of any object at all—the possibility of their changing and even going out of existence.

It is worth repetition since it is fundamental to my case: all and any structures are constrained by the mathematics of possibility, potential, and change that make them available not only as causally investigatable but as manipulable into novel objects by imaginative and purposeful teleological agents like ourselves.

Natural and human, organic and inorganic structures thus all share the basic necessary traits that define them as legitimate ontological inhabitants of anything making up a world: structure, aesthetics, and modal possibility, constrained by the mathematical functions they exemplify.

The overarching picture provided by this metaphysical trinity underpins the manifestation of anything that might count as part of the "physical"

23. I mean to include here the properties of the quantum world which cannot be said to be "physical" in the traditional common-sense view of the world nor in the Newtonian view. "Physical" here is really a shorthand for what is taken by physics to constitute a necessary basis for all other "natural" structures.

world, inclusive of that part of it which we are inclined to call "nature." The tools of our epistemological engagement with nature, both abstract and concrete, are attempts to develop an understanding of how the structures of nature work. "Methodological" naturalism helps define the means and which contingent properties that can yield that understanding. It does not rule out the broader picture.

Deacon's Incomplete Nature and Descartes: Opposition or Similarity?

I now return to the work of Deacon. With the picture above in mind, I hope to show that even a sophisticated grasp of teleology, the biological world, and the physics of energy that Deacon displays even while opposing Cartesian mechanism in biology can still unwittingly accept it for the other natural sciences.

Unlike many philosophers who hope to resolve the tension between teleology and physical causal aim by finding a suitable philosophical relation between the two domains that preserve the ontological primacy of the latter, Deacon acknowledges that any resolution of this issue must start from the assumption that, as the title of his extensive book and research on the question indicates, "nature is incomplete."[24]

The title of the book summarizes his thesis, which is that biological science, as a branch of nature, cannot avoid the empirical pressure placed upon it by biological data to incorporate explanations of "purpose" into its theories. He thinks that what makes this a theoretically unpalatable task for some is a prior implicit attachment to a Cartesian derived mechanistic view of nature. Deacon proposes that it is "the tacit importation of a human artifact view of the world, with its implicit design like logic into a materialist metaphysics that restricts the introduction of anything like final causal relationships . . . where design and purpose can only be imposed from the outside."[25] By the "outside," he means by "rational human persons" who have mastery over the teleological concepts that "design" involves.

"Incomplete nature" is intended to signify that rather than trying to squeeze the teleological into the physical, our conception of physical and chemical processes and the causal understanding (that of efficient causation) that accompany them need to expand when dealing with organic life. Within the later term he wants to include sentient biological creatures of all kinds. Deacon's essential idea for doing this is to revise the principle of

24. Deacon, *Incomplete Nature.*

25. Deacon, *Incomplete Nature*, 37.

"efficient" and "mechanical" causation in the light of the "fundamental requirements of life from the point of view of physics."[26]

Because he thinks that the terms "teleology" and "intentionality" have "mentalistic connotations,"[27] he proposes a new term to separate out the intrinsic mentality of teleology and intentionality which he calls "ententionality." He thinks that his new term encompasses a notion of teleology that includes and explains human teleology in its full flourishing when mentality, or consciousness, is "added" to it.

He thinks that the idea of "incompleteness" as a motif for reintroducing teleology into nature provides him with a technical opening which can help domesticate the concepts of "purpose" and "final causation." The problem for which he wants to find a theoretical solution within biology is one that he thinks is hindered by its roots in Cartesian dualism:

> The metaphysical problem of reintegrating purposiveness and subjectivity into theories of physical processes [has] led many thinkers to propose a kind of forced marriage between mental and physical modes of explanation.[28]

He extends his critique of the "forced marriage"—in essence, the mind–body problem bequeathed by Descartes—to include theories which address the dynamics of complex systems, including Darwinian processes. Despite the latter, for example, yielding

> enormous advances in our understanding of the orderliness observed in living, neurological and even social processes . . . dynamical systems . . . are often forced to explain away the end-directed and normative characteristics of organisms because they implicitly assume that all causally relevant phenomena must be instantiated by some physical substrate or energetic difference. Consequently, they are limited in their power to deal with the representational features of mind, as are simple mechanistic accounts.[29]

Similarly, he also critiques theories of mind that rely on concepts such as information, representation, and function, and assumes these terms are

26. Dennett, "Aching Voids and Making Voids," 322.

27. Deacon, *Incomplete Nature*, 25.

28. Deacon, *Incomplete Nature*, 5.

29. Deacon, *Incomplete Nature*, 5.

> shorthand proxies . . . for otherwise well understood, but long, clumsy and arduous mechanistic accounts; that is accounts involving only the extended physical properties of things.[30]

"Ententionality" and teleology, he argues, are phenomena that emerge with the neurobiology and behavior of organic structures and creatures. It is intrinsic to his case that "goals," "purposes," and "meanings" are also emergent phenomena, nowhere present in the universe until their appearance with certain forms of evolved life.

I hope it does not do Deacon an injustice to summarize his view as the idea that we must somehow find a theoretical way of importing the language of teleology into our understanding of the behavior of minded organisms. This must be accomplished in such a way, he proposes, as to not only explain the neurobiological origins of our capacity to think teleologically and normatively, but also the *origins* of teleological concepts and reasoning themselves.

His account of how purposeful behavior emerges from matter is highly technical, involving an attempt to develop structures based on their need for "energy" as understood in the second law of thermodynamics, particularly mechanisms for storing and generating their own energy. These structures, in the end, must develop ways of obtaining "sources" of energy. It is out of these requirements that the "foundations for telodynamic life are laid."[31] The essential point is that the appearance of teleology and normativity in the world are, for Deacon, an empirical and contingent issue.

That this is his view or something close to it is supported by a paragraph during the course of which he discusses primeval earth prior to the evolution of the hominids. I quote this in full because it is a common picture of the way that those scientists who see the need to admit teleology frame the problem:

> It can hardly be doubted that there was a time when these phenomena (the sorts of processes and relationships that involve purposes, goals, meanings, and the like) were not in evidence on the surface of the earth, although they are everywhere in evidence at present. Certainly, none were present when the earth's surface was still hot enough to boil water. . . . They probably did not just drop in from anywhere, though if they did, their origin would need to be explained there. Wherever and whenever they

30. Deacon, *Incomplete Nature*, 5.

31. Dennett, "Aching Voids and Making Voids," 322.

> arose, they were new, even if they are now mundanely common in our experience.[32]

The metaphysical conclusion to draw from this segment is that teleology, goals, and purposes could not be said to "exist" in any sense of the word prior to the evolution of sentient organic life: Deacon's position is not, as I understand it, that sentient creatures emerge into a world where teleological concepts, final purposes, and reasons become *available* to them as they become more complex creatures, but that they *construct* the necessary frameworks themselves. "Teleology," final "purposes," and so on are entirely emergent cognitive constructions of biological tissue (or some aspect of a CNS and the behavior it governs).

There are two essential problems with this argument. Firstly, if no other scientific discipline admits "purposes" as final causes because no other discipline has any space in a conception of how things work for anything other than the progression of causes, then how the "purposes" which emerge with biology gain any traction in the rest of the physical world is a problem.

What is it about the physical world so conceived, in other words, that can make it hospitable to *rational* teleological manipulation? This is the problem I have addressed at length in the earlier section. The problem is a qualitative leap from a purely physical description to "reason," "purpose," "intentionality," and value-governed agency. It is as if someone were to fully explicate all the movements and pieces on a chessboard in terms that never referred to rules and at some point in that explication declare, ". . . and those are the rules of chess."

The second problem is that *how* it is possible that creatures can become responsive to normative and teleological criteria does not provide an answer to *where* in the world normative standards are to be found.

To explain (if he can) how an organism becomes able to recognize and respond to normative criteria is not the same thing as explaining from whence those standards derive and their proper place in the metaphysical scheme of things. To show what may be required cognitively (or neurobiologically) to play a game of chess is not the same thing as understanding the rules that define what that game is (or *what* a rule is for that matter). One response might be that such considerations are unnecessary since evolution takes care of that. Creatures don't follow rules in the sense a rational mind does, but simply produce behavior that has become learned for survival. There is much obviously to be said in favor of this. However, there is a deeper metaphysical problem. Why should survival be built into the natural world? No doubt it is there, but why should it be patterned into the

32. Deacon, *Incomplete Nature*, 38.

character of a world that creatures can evolve, for whom an ultimate goal is survival? A final purpose, surely, and if so, a teleological property built into the very nature of how one is to understand what a natural world *is*. A rule, one might want to say, defining life. Secondly—and here I cannot deal comprehensively with so vast a topic—intelligent self-aware rationality has a different character. Rational minds grasp that they are answerable to normative standards in the world that do not emerge from themselves but that bind their behavior, practically, and in the knowledge-seeking activities of the sciences.

Normativity in the World and Thinking Normatively

It is no answer to say that we can construct or invent the idea of a standard or rule, since to construct a standard or use a rule presupposes that one already has an idea of what a standard *is*. To have an idea of what a standard is, is *already* to be in possession of the idea of some judgment being answerable to something other than itself. To be already in possession of the idea of standards suggests that human beings come into cognitive *contact* with a world of teleology and normativity, rather than build it.

It is the same mistake, for example, of conflating the "laws" of logic or mathematics with what is a necessary but insufficient condition of being able to understand them. Additionally, one cannot move from any set of properties described in terms of the regularities of any set of causal relations to the realms of "logic," "mathematics," and "representation." The move from the physical to the mathematical and logical is a conceptual one. It is a conceptual framework that describes the conditions of existence without which the physical world and its inanimate or animate structures and functions is logically impossible.

It may be, as Deacon suggests, that cognitive neurobiological attainments of varied and multiple sorts are a necessary condition of grasping logical and normative truths. But these truths are, as it were, present in the universe as metaphysical conditions on the *possible* existence of any rational and sentient beings.

Deacon makes the same mistake as Descartes, which is to fail to recognize that there *is* no physical world without the logical, mathematical, structural, and hence normative truths which give existential shape to physical structures. These aspects of the world are those we emerge into contact with given various necessary, but not sufficient, complex evolutionary and neurobiological preliminaries. It is also the aspects of the world that structure the conditions of existence of any living creatures (or inanimate structures, for that matter).

A "Nature" Full of Aesthetically Mathematical Structures

Once we have seen this point, then the connection between physical items taking some structure, mathematical descriptions as determining the possible and actual form any such physical things can take, and the aesthetic features that they thereby manifest becomes visible. They provide the constraints on the intelligibility of *any* structure manifesting itself as a feature of the natural world, whether the arcane world of particle physics or the biology and patterns of behavior in which Deacon has his primary interests. In fact, the underpinnings of "neurobiology" in which Deacon places his hope are themselves the outcome of the structural possibilities of any world such as we have it.

Deacon does not really have a theory of artifacts either. He recounts the traditional picture of them as projections of human interests onto blank physicality.

"Nature" understood through the lens of "structure," "mathematics," "aesthetics," and the relations between them, on the other hand, has precisely the place for ontological equivalence that Baker Rudder proposes for both physical objects and processes *and* those that are the outcome of intentional human production.

Any "object," "process," "event," or substance which comes into physical being must be intelligible in virtue of these features, whether the product of non-sentient causes, "simple" sentience, or fully teleological rationality such as we possess.

Once we lift the restrictive scope through which ontological naturalism views the world by a mistaken leap from the causal restrictions of methodological naturalism to the "personal"/"impersonal" distinction it imposes on metaphysics, the enchantment of "nature" is restored: aesthetic properties are manifested in impersonal objects, and the artifacts of personal creation obey the same underpinning unifying principles as those of "natural" objects.

Nature can thus exhibit beauty in its constructs enabled by the mathematical aesthetics of structures. We are similarly provided by "nature" not only with the material for the construction of artifacts but also access to the aesthetic, mathematical, and normative structures of the empirical that enable *us* to design many things, including things of great beauty.[33]

33. It comes along with the ability to distinguish beauty and ugliness that sometimes

However this is finally to be understood in the theories of the biological and zoological sciences, it contributes to the evidence that nature has teleological potential within its ontological resources. That potential is expressed in its fullest form by human persons but only because "nature" can accommodate both us and the principles of purposeful behavior.[34]

Teleology in Nature and Divine Provision

Once we grant the necessity of normativity in deep metaphysical nature, the door is open to the idea that for the physical world to finally issue in a sentient nature, basic physical structures must have carried in their metaphysical heart the potential for conscious, rational teleological creatures to emerge.

It is a natural response to the recognition of the coherence of that intricate metaphysical structure to ask what might explain its origin. If that final existential resting place is taken to be "impersonal" or natural in the ontological sense that excludes teleology and the personal, then we are faced with the problems with which this chapter has been concerned: the isolation of teleology, and human artifacts from a broader necessary picture of the world, and the inability of a biological science to answer the question of normative provenance.

An origin, however, that *includes* teleology and the personal seems better fitted for the purposes of articulating the intelligible coherence of a "working" physical world, its careful a priori underpinnings, and their equally intelligible relations to teleology, aesthetics, beauty, and mathematics manifest in artifactual and "physical" objects and events.

That origin, I postulate, is a personal, divine, and teleological one, and it is the purpose of the next chapter to elucidate beauty, the nature of its transcendent power, its author, and how the human capacity for artifactual construction echoes the divine artifact that is the natural world.

---

the latter are sadly chosen.

34. It is worth mentioning that the very idea of "evidence," the history of evidence, and the history of science is dependent on normative (and cultural) prescriptions for its existence. Is it at all intelligible, for example, that the history of science relying as it does on teleological concepts and the presence of thinking persons and their artifacts might turn out to be nothing other than a conceptual framework reducible itself to the "impersonal" causal forces of the natural sciences? The incomprehension of this possibility is, I suggest, manifested in its posing.

Chapter 4

Universal Structures of Beauty and Their Transcendental Teleology

SUMMARY

HAVING ESTABLISHED NORMATIVE PROPERTIES as metaphysically fundamental in both the personal and impersonal worlds—particularly in the case of aesthetics—this chapter turns to the question of beauty: its apparent universality, its distinctive experiential and structural features, and the role it may play in arguments concerning divine existence. The underlying issue is whether beauty, as it appears across domains as diverse as mathematical physics, the natural world, and human artifact, can be understood as a unified phenomenon, and whether such unity bears upon the intelligibility of divine authorship.

OVERVIEW OF CHAPTER

If it is correct, as I have argued, that everything which exists must possess an aesthetic character, then it follows that among the very different events and objects that constitute the world, many will be beautiful. Beauty, on this view, is not an anomalous or exceptional feature but one that arises naturally from the conditions under which anything can be structured at all. It would therefore be surprising if beauty were rare; on the contrary, it should be expected to appear across a wide range of domains, albeit in different forms and degrees.

Experiences of beauty—whether in the physical sciences, in the natural world, or in human artifacts—are notable for the kinds of responses they tend to provoke. They frequently give rise to existential questioning and to forms of longing that seem to exceed the immediate context in which they arise. What is striking is that such responses are not confined to those already disposed to theological reflection. They are equally reported by those with no prior commitment to theism. Something in the nature of beauty appears to exert a pressure on thought, drawing it beyond what is simply presented, and inviting reflection on what, if anything, might underlie such experience.

For this reason, beauty would appear to constitute a natural point of entry for natural theology. Yet it has not, in fact, occupied a central role in such discussions. One reason for this may lie in the way the appeal to beauty is often interpreted. The suggestion that beauty produces transcendental or quasi-religious responses is frequently taken to indicate nothing more than a feature of subjective experience. On this view, the force of such experiences is explained in terms of human psychology: they are affective responses to certain configurations of stimuli rather than disclosures of anything about the nature of the world itself.

Underlying this interpretation is a more general assumption, namely that subjective modes of apprehension and objective features of reality are sharply distinct and that only the latter carry ontological significance. If this division is granted, then aesthetic experience can be safely confined to the domain of the merely personal, without consequence for our understanding of the structure of reality.

Against this, I suggest that once the broader framework developed in earlier chapters is brought into play—particularly the relation between structure, artifact, and aesthetic character—the experience of beauty begins to look rather different. It no longer appears as an incidental accompaniment to perception but as one of the ways in which structure is disclosed. In particular, the conjunction of structure, artifact, and aesthetic evaluation has a tendency to draw one toward the idea of personal intent, not as an imposed hypothesis but as something that becomes intelligible within the experience itself.

The comparison with artifacts is crucial. In the case of human creations, we do not infer purposiveness from external evidence; rather, we recognize it in the form of the object itself. To see something as an artifact is already to see it as expressive of intention. If, as I have argued, natural objects share in the same ontological conditions as artifacts—if both are necessarily aesthetic in their constitution—then the strict opposition between the personal and the natural begins to weaken. The idea that objects of nature cannot display, as part of their identity, the personal hand of creation becomes less secure.

This weakening is not merely suggestive but structurally significant. If teleology and normativity can be shown to belong to the conditions under which physical structures exist, then such structures fall within the scope of teleological description. Once this is granted, the move to divine intentionality is not a leap beyond the argument but a continuation of it. The language of divine artistry arises at this point not as metaphor but as an attempt to articulate a mode of understanding capable of encompassing both the personal and the physical within a single ontological framework. Human artifacts already instantiate teleology; if structurally analogous forms of beauty are found at the foundations of the natural world, it becomes a natural—indeed almost spontaneous—inclination to interpret such structures as expressive of a form of intentional creation.

The Logic of Aesthetic Universality and Aesthetic Empiricism

I have maintained that aesthetic properties are logically prior to the coming into being of any structured element of the universe. The form of this argument can be restated more explicitly as follows: Let *x* denote any aesthetic property and *y* denote any structure. If *x* is a necessary constituent of something being *y*, then the absence of *x* entails the absence of *y*. In other words, if there were no aesthetic character whatsoever, there could be no structure; and if there is no structure, there is nothing that can be said to exist in any intelligible sense. The conclusion, therefore, is not merely that aesthetic properties accompany structures but that they are constitutive of them.

It is important, however, not to overstate what follows from this claim. The argument does not entail that any given aesthetic property must be instantiated in every structure, nor that instances of aesthetic properties are identical with instances of structure. Rather, it establishes a condition of possibility: for anything to exist as a structured entity, it must possess *some* aesthetic determination. What is ruled out is not variation but the complete absence of aesthetic character. Thus, while all structures are necessarily aesthetic, they need not be aesthetically identical.

This distinction between necessity and variation allows us to preserve both the universality of aesthetic character and the diversity of its manifestations. It follows that although we can know a priori that every structure must have an aesthetic aspect, we cannot determine *which* aesthetic qualities any particular object possesses independently of experience. Whether something is beautiful, harmonious, unified, or otherwise must be established through encounter with it. Aesthetic knowledge, in this sense, is both grounded in necessity and dependent on experience: it is at once a priori in its general form and a posteriori in its particular application.

Given that every structure must be aesthetically qualified, it becomes a reasonable expectation that a wide range of aesthetic qualities will be instantiated across the world. Among these, beauty will appear in multiple domains: in mathematical structures, in works of art, in natural forms, and in the character and activity of human persons. The extension of aesthetic predicates across such a wide range of objects is not accidental but reflects the generality of the conditions under which structure is possible. As Patrick Sherry observes, the term "beautiful" is not confined to material things but is also applied to ideas and scientific theories, indicating that its scope is far broader than might initially be supposed.[1]

This breadth of application immediately raises a question that will run throughout this chapter: What is it that we are identifying when we call something "beautiful"? If beauty can be predicated of such diverse entities, is it a single property, a family of related properties, or simply a convenient label for a range of otherwise disparate experiences? More specifically, is beauty something that belongs to the object, something that arises in the subject, or something that emerges from the relation between the two? The way in which we answer these questions will determine whether beauty can bear the metaphysical and theological significance that I wish to assign to it.

PART 1

Objectivity, Structure, and Aesthetic Experience

Within the analytic tradition, there has often been a strong inclination to suppose that meaningful discourse about a property or object requires that it be definable, preferably in precise and formally specifiable terms. Whether such definitions are always attainable, and whether intelligible discourse is possible in their absence, is itself a matter of ongoing debate. In what follows, I adopt the view that many concepts—particularly those relating to aesthetics—do not admit of strict identity conditions that neatly divide instances from non-instances. This does not render them unintelligible; rather, it suggests that their intelligibility operates under different conditions than those governing more tightly circumscribed concepts.

The work of Frank Sibley is instructive in this regard. Sibley observed that aesthetic terms span a wide range of uses and can be grouped

1. Sherry, *Spirit and Beauty*, 35–36.

into different kinds.[2] Some terms, such as "dynamic" or "tightly knit," are not exclusively aesthetic but may be pressed into aesthetic service, while others—"beautiful," "graceful," "elegant"—are primarily aesthetic in their function. This distinction helps to clarify the diversity of aesthetic language, but for present purposes it is useful to introduce a further category, which I will call *explanatory descriptors.*

Explanatory descriptors are those structural features that are invoked in accounting for why something is judged to possess aesthetic merit. Terms such as "simple," "unifying," "proportionate," "harmonious," and "symmetrical" belong to this class. They refer to aspects of the organization of a structure—relations between parts and wholes, patterns of coherence and balance—which are frequently cited in explanation of aesthetic judgment. Importantly, these descriptors are not confined to any single domain. They are employed in discussions of human artifacts, natural phenomena, and scientific theories alike, suggesting that they track features of structure that are general rather than domain-specific.

In many cases, when we attempt to explain why something is beautiful, we appeal—explicitly or implicitly—to these kinds of structural features. We say that a piece of music is beautiful because of the balance between its themes, that a mathematical proof is elegant because of its simplicity, or that a building is graceful because of the proportional relations between its elements. Such explanations rarely purport to reduce beauty to these features; rather, they indicate aspects of the structure that help to make intelligible the aesthetic response.

Within the theological and philosophical tradition, one of the most well-known attempts to bring these elements together into a unified account of beauty is that of Thomas Aquinas. As discussed by commentators such as Christopher Scott Sevier[3] and Umberto Eco,[4] Aquinas offers a conception of beauty that integrates both objective and subjective components. On the one hand, beauty is grounded in the intrinsic properties of things—integrity, proportion, and clarity; on the other, its appreciation requires a perceiver suitably equipped to recognize and respond to those properties.

This dual aspect is of particular importance for the present argument. It allows us to maintain that beauty is neither wholly subjective nor wholly independent of experience. Rather, it is rooted in the structure of things while being disclosed through a form of cognitive and perceptual engagement. The experience of beauty is thus not an arbitrary projection but a

2. See Benson et al., *Frank Sibley.*

3. Sevier, *Aquinas on Beauty.*

4. Eco, *Aesthetics of Thomas Aquinas.*

response to features that are genuinely present in the object, even if those features are not exhaustively captured by the experience itself.

This account aligns closely with the earlier claim that there are two modes of access to the aesthetic structures of mathematical physics: one through formal description and empirical investigation, and the other through a form of cognitive resonance with the structure's aesthetic qualities. The latter is possible because the structure itself possesses properties—unity, symmetry, simplicity—that are capable of eliciting such a response. In this way, aesthetic experience becomes a mode of apprehending aspects of structure that are also accessible, in different terms, through formal analysis.

As R. B. Bernier interprets Aquinas, beauty is closely linked to knowledge: "Beauty has to do with knowledge, and we call a thing beautiful when it pleases the eye of the beholder. . . . Beauty is a matter of right proportion for the senses delight in rightly proportioned things. . . . Things that are lacking in integrity or perfection are for this reason ugly."[5] The notion of "right proportion" is central here, not only as a structural feature but as a point at which objectivity and experience intersect. It provides a way of understanding how beauty can be both grounded in the nature of things and apprehended through perception, without collapsing into either pure subjectivity or purely formal description.

Beauty, Knowledge, and the Integrity of Structure

The association between beauty and truth, or between beauty and the "rightness" of what expresses it, has a long history, as discussed in chapter 2. In contemporary usage, this connection has often weakened, but it remains implicit in many contexts. Catherine Pickstock, reflecting on the transformation of aesthetic understanding in the modern period, notes that a decisive shift occurred with the separation of the so-called transcendentals—being, unity, truth, goodness, and beauty—which had previously been regarded, from Plato through Aquinas, as mutually convertible.[6] Within that earlier framework, beauty was not an isolated property but one aspect of a unified conception of reality.

Classical thought, particularly in its Platonic and neo-Platonic forms, understood truth in terms of the proper ordering of things within a hierarchy

5. Bernier, "After Aquinas," 96. Whether there are exegetical difficulties associated with Bernier's interpretation and translation is not important for my purposes here. The quotation is sufficiently close to a common understanding of the relation between form and aesthetic perfection, or beauty.

6. Pickstock, introduction to Gill, *Beauty Looks After Itself*, iii.

of being. Beauty, on this view, is encountered when something occupies its appropriate place within that order, exhibiting a form that corresponds to its nature. While it is not necessary to adopt the full metaphysical apparatus of Platonism, the underlying insight can be preserved in a more modest form. Empirically, it is often the case that aesthetic merit is associated with the sense that a structure has achieved a kind of optimal realization—that it could not have been better arranged given what it is.

To describe something as beautiful, in this sense, is to suggest that it exhibits a form of structural integrity: that its elements are arranged in such a way as to produce a sense of completeness or finality. This is not merely a matter of subjective preference but of an experienced fit between parts and whole, between function and form. The closer a structure comes to this condition, the more readily it is judged to be beautiful.

This connection is evident in discussions of artifacts. No less a figure than the Enlightenment Scottish philosopher David Hume (1711–1776), in *A Treatise of Human Nature*, observes of the beauty of many things,

> scritoires, chimneys, coaches, and saddles . . . [that it is] "a universal rule" . . . that their beauty is chiefly deriv'd from their utility, and from their fitness for that purpose, to which they are destined.[7]

Hume's intention was to argue that beauty is derived from utility, and thus ultimately from human response. However, the observation itself can be interpreted differently: it suggests that aesthetic evaluation is closely tied to the structural integrity of the object, including its functional coherence.[8]

It is important, however, not to reduce aesthetic judgment to utility. While functional adequacy may contribute to aesthetic value, it does not exhaust it. Explanatory descriptors—such as symmetry, unity, and proportion—are often invoked not to reduce beauty to function, but to illuminate aspects of the structure that contribute to the experience. The meter of a poem, for example, may be analyzed to explain how it produces a certain rhythm that supports its expressive content; the analysis does not eliminate the aesthetic experience but helps to make it intelligible.

Similarly, in historical writing, a work may be praised for the integrity with which it weaves together disparate events into a coherent narrative. In such cases, the aesthetic judgment reflects an appreciation of structural organization rather than a simple response to utility. Comparable observations

7. Hume, *Treatise of Human Nature* 2.2.5 (364).

8. Parsons and Carlson devoted a book to the question of "functional beauty," its history, and its relation not only to human artifacts but also to the natural world. See Parsons and Carlson, *Functional Beauty*.

can be made in relation to music, painting, and other art forms, as well as in the natural sciences, where descriptors such as "symmetry," "unity," and "harmony" play a role in theory evaluation.

In the case of mathematical physics, this relation becomes particularly pronounced. As I previously discussed, physicists such as Dirac have explicitly appealed to aesthetic criteria in the development and assessment of theories. As I noted, for Dirac, the beauty of a mathematical structure—its simplicity, elegance, and internal coherence—was not merely an optional embellishment but a guide to its truth. The suggestion here is that the structural features that give rise to aesthetic appreciation are also those that underpin theoretical success.

At this point, the attempt to treat aesthetic properties as merely subjective begins to look increasingly strained. While it is true that our responses to aesthetic features depend on our cognitive and perceptual constitution, this dependence is not unique to aesthetics. All forms of knowledge are mediated by our capacities, yet this does not lead us to conclude that the objects of knowledge are therefore unreal or merely projected. To argue that aesthetic properties are subjective in this sense is to rely on a general skepticism that would apply equally to other domains.

If aesthetic properties are necessary constituents of structure, then they will be present wherever there is any intelligible grasp of reality. It is therefore to be expected that our cognitive, affective, and even neural systems will be attuned to them. Aesthetic experience, on this view, is not an optional addition to our engagement with the world, but one of the ways in which the world's structure becomes accessible to us.

At the same time, beauty is often sought for its own sake and is experienced as possessing a kind of transcendence. It is not merely useful nor is it confined to practical concerns. This conjunction—of structural grounding and transcendental effect—will become increasingly important as the argument proceeds.

The Universality of Beauty and the Particularity of Beauty in Physics

The claim that beauty is both widespread and structurally grounded gives rise to an immediate difficulty. If beauty is found across such a wide range of objects—mathematical structures, natural forms, works of art, and human character—how is it to be understood as a unified phenomenon? More specifically, if beauty is associated with certain structural features, such as symmetry, proportion, and harmony, how can this be reconciled with the fact that objects possessing very different configurations, and sometimes lacking these features altogether, are nevertheless judged to be beautiful? The problem can be sharpened further by considering the apparent divergence between the use of aesthetic concepts in mathematical physics and their use in the human world. If the term "beauty" functions differently in these domains, then its apparent universality may be more linguistic than ontological.

A related concern arises from the variability of aesthetic judgment in human affairs. It is a commonplace observation that aesthetic responses differ across cultures, historical periods, and individuals. What is regarded as beautiful in one context may be indifferent or even displeasing in another. If this variability is taken to indicate that aesthetic judgment is fundamentally relative, then the claim that beauty in the sciences reflects objective features of structure becomes difficult to sustain. It may be argued that the appeal to beauty in physics is itself a contingent feature of a particular intellectual tradition—perhaps a residual inheritance from earlier metaphysical frameworks—and that future developments may dispense with such language altogether.

These considerations, taken together, threaten to undermine the connection I wish to establish between beauty in the human and physical domains. If the beauty of mathematical physics is of a fundamentally different kind from that of human artifact and experience, then the inference from one to the other becomes unstable. In particular, the suggestion that the presence of beauty in both domains points toward a common origin—ultimately toward divine intentionality—loses its force if the term "beauty" does not retain a sufficiently unified meaning.

Any adequate response must begin by acknowledging that there are genuine differences between the use of aesthetic concepts in these domains. In mathematical physics, aesthetic judgment is often closely tied to a relatively small set of structural features—simplicity, symmetry, generality, and unity—which are regarded as indicators of theoretical virtue. As has often been noted, physicists such as Dirac were willing to privilege such criteria to a remarkable degree, treating them as guides to truth rather than as mere heuristic preferences. Jane McDonnell observes that mathematical beauty is "of a completely different kind" from the aesthetic responses associated with the arts. "Mathematical beauty is the same in all countries and at all periods of time & advances with a timeless beauty linked to ideas of simplicity, symmetry, generality and truth."[9]

9. McDonnell, *Pythagorean World*, 19.

By contrast, aesthetic judgment in the human world appears far more variable. The appreciation of a work of art, a piece of music, or a literary text may depend on a wide range of factors, including cultural background, education, and personal experience. It is therefore tempting to conclude that the apparent objectivity of aesthetic judgment in physics is not matched in the human domain, and that the two uses of "beauty" are only loosely related.

However, this conclusion does not follow as directly as it might first appear. The variability of aesthetic judgment in human affairs does not in itself entail that such judgments are purely subjective or devoid of objective reference. Differences in response may reflect differences in familiarity, training, or sensitivity to the relevant features of the object, rather than the absence of such features. The difficulty of appreciating a complex piece of music, for example, does not imply that its aesthetic qualities are unreal, but rather that their apprehension may require a certain degree of attunement or understanding. The same point applies, *mutatis mutandis*, to the appreciation of mathematical structures.

Moreover, the structural features identified in the sciences are not wholly absent from the human domain. Concepts such as symmetry, balance, proportion, and unity play a central role in many forms of artistic and cultural production, even if they are not always foregrounded in the same way. The apparent divergence between the two domains may therefore be less a matter of fundamentally different kinds of beauty than of differences in emphasis and context. In mathematical physics, certain structural features are especially salient because of their relation to theoretical explanation and empirical adequacy; in the human world, a wider range of factors may be involved, leading to greater variability in judgment.

It is also worth noting that aesthetic judgments, in both domains, are typically expressed in a form that purports to describe the object rather than merely report a subjective response. When someone claims that a theory is elegant or that a piece of music is beautiful, the claim is usually intended as a statement about the object itself, even if it is recognized that others may disagree. This feature of aesthetic discourse aligns it more closely with factual and moral discourse than with the mere expression of preference. While this does not by itself establish the objectivity of aesthetic properties, it does indicate that our ordinary practices of judgment treat them as if they were features of the world rather than projections onto it.

Disagreement, in this context, does not necessarily undermine objectivity. As in the case of moral judgment, it may instead reflect the complexity of the phenomena under consideration and the difficulty of achieving agreement about them. The possibility of error—of failing to recognize beauty where it is present, or of attributing it where it is absent—is built into the practice of aesthetic evaluation. This would be unintelligible if aesthetic judgments were understood as nothing more than expressions of individual preference.

The apparent tension between universality and variability can therefore be reinterpreted. Rather than indicating that beauty lacks a coherent core, the wide range of its manifestations may point to the depth and complexity of the aesthetic dimension of reality. The fact that beauty can be encountered in such diverse forms—while still exhibiting certain recurrent features—suggests that it is not tied to a single set of criteria but is nevertheless grounded in the general conditions of structure. The variability of aesthetic judgment in the human domain may thus be understood not as evidence of its subjectivity but as a reflection of the richness of the forms in which aesthetic character can be realized.

At the same time, the relative stability of aesthetic judgment in mathematical physics does require explanation. It may be that the domain itself imposes constraints that limit the range of relevant features, making certain forms of organization—such as symmetry and simplicity—particularly significant. In this case, the apparent objectivity of aesthetic judgment in physics would not be a sign of its uniqueness but of the more restricted conditions under which aesthetic evaluation operates in that domain.

If this is correct, then the difference between the human and physical cases does not preclude a common underlying phenomenon. Rather, it suggests that beauty may be instantiated under different conditions, with varying degrees of determinacy. This opens the possibility that the same basic feature—an aesthetic dimension grounded in structure—may be present in both domains, even if it is manifested in different ways. If so, the inference from the beauty of human artifacts to the beauty of natural structures, and from there to questions of intentionality, remains intact.

The Criteria of Identity for Concept Membership

The preceding discussion brings into focus a more general issue concerning the nature of concepts and the conditions under which they can be said to apply. If beauty is to function as a unifying concept across such diverse domains, it cannot depend on a rigid set of criteria that must be satisfied in every instance. The expectation that a concept must be definable in terms of necessary and sufficient conditions is, in many cases, too restrictive to capture the way in which our conceptual practices actually operate.

This point is forcefully illustrated in the work of the philosopher Ludwig Wittgenstein, who argued that many of our concepts are better understood in terms of what he called "family resemblances." Rather than sharing a single defining feature, the members of a category may be related by a network of overlapping similarities. Wittgenstein famously compares this to a rope composed of many strands, none of which runs through the entire length but which together form a continuous and functional whole.[10]

This insight is developed empirically in the work of the cognitive psychologist Eleanor Rosch.[11] According to Rosch, categories are structured in terms of typical examples, and membership is determined by degrees of similarity to these prototypes rather than by strict adherence to a set of criteria. Some instances may be more central than others, but all may still belong to the category in virtue of their relation to the prototype.

Applied to the concept of beauty, this approach allows us to reconcile diversity of application with conceptual unity. Different domains may emphasize different features—symmetry and simplicity in physics,[12] perhaps expressiveness or complexity in art—while still being related through a network of shared characteristics. The absence of a single defining criterion does not entail that the concept is empty or incoherent; rather, it reflects the richness of the phenomena to which it applies.

This also helps to explain the coexistence of variability and stability in aesthetic judgment. In domains where the relevant features are tightly constrained, as in certain areas of mathematical physics, aesthetic judgments may exhibit a high degree of convergence. In other domains, where the range of relevant features is broader and more context-dependent, greater variability may be expected. In both cases, however, the concept of beauty

10. A good introduction to Wittgenstein's later work, *Philosophical Investigations*, in which this case is made, is provided by Schroeder, *Wittgenstein*. Three chapters in Glock, *Wittgenstein*, examine issues relevant here in greater depth: Ammereller, "Wittgenstein on Intentionality"; Rundle, "Meaning and Understanding"; and Arrington, "Following a Rule." The literature on Wittgenstein is vast, and I have simplified his position to the barest acceptable minimum.

11. Rosch has been a central figure in cognitive psychology and has generated an extensive secondary literature. I include here two of her seminal papers: Rosch, "Natural Categories," and Rosch, "Cognitive Reference Points."

12. An empirical paper by de Regt examines the criteria mathematical physicists employ in theory evaluation. These appear largely to be a small number already discussed in this work, including symmetry, simplicity, and "conformity with the whole." See de Regt, "Beauty in Physical Science." McAllister similarly argues that scientists tend to employ criteria drawn from a "classical, formalist aesthetic." He suggests, however, that a "different aesthetic" valuing "differentiation" and "complexity" may better represent the full range of scientific activity. See McAllister, "Recent Work on Aesthetics." If my argument is correct, however, there remains ample space for aesthetic values of these kinds, though I do not regard them as fundamentally different but rather as issuing from the same broad aesthetic field that beauty itself encompasses.

may still be doing the same kind of work, even if the criteria for its application differ.

The implication for the present argument is that the lack of a single, fixed set of criteria for beauty does not undermine its role as a unifying concept. On the contrary, it allows for the possibility that beauty may be instantiated in a wide range of forms while still retaining a coherent identity. This, in turn, supports the claim that the beauty encountered in different domains—human, natural, and scientific—may be understood as manifestations of a single underlying feature of structured reality.

Beauty: Simplicity and the Transcendence of Structure

If the preceding discussion has emphasized the relation between beauty and structure—particularly through explanatory descriptors such as symmetry, unity, and proportion—it is equally important to recognize that beauty cannot be adequately captured as a mere function of these features. There is, alongside its structural articulation, a striking simplicity to beauty that resists reduction to the sum of its parts. This simplicity is not to be confused with the simplicity of a minimal structure nor with the absence of complexity. Rather, it is a feature of the experience itself: beauty presents itself as something immediately apprehended, even where the structure that gives rise to it is intricate and demands extended engagement.

This immediacy has been well captured by David Bentley Hart, who notes that the apprehension of beauty is characterized by a certain simplicity in the act of perception.[13] By this, he does not mean that beauty is always recognized instantaneously or without effort but that, in any given aesthetic experience, the judgment of beauty is not mediated by external considerations. It is not derived from utility, nor from moral evaluation, nor from theoretical reasoning. Rather, it is encountered as something that presents itself as worthy of attention in its own right. In this sense, beauty is not inferred but given.

At the same time, this simplicity is accompanied by a peculiar tendency of aesthetic experience to exceed the object that occasions it. Beauty does not merely present itself as a property of a thing; it often seems to point beyond the thing, to evoke a sense of something more than what is immediately before us. This is evident in the way aesthetic experience can give rise to feelings of elevation, longing, or even dislocation from the ordinary frame of reference within which objects are usually encountered. The object remains present, but the experience is not exhausted by it.

13. Hart, "Beauty, Being, Kenosis," 246.

A similar point is made by D. C. Schindler, who describes beauty as engaging the whole of the person, having an

> astonishingly comprehensive quality . . . it appeals to the whole of us . . . engages . . . our mind and senses at once . . . enlisting them in the common project of *perceiving* beauty. . . . Beauty thus involves our sense of transcendence, our sense of being elevated beyond ourselves . . . at the very same time it appeals to our flesh, our most basic natural instincts. In this way it reaches into the depths of our nature, touching what is most basic and essential.[14]

Beauty, on this account, is not confined to a particular faculty but involves a kind of total engagement in which perception, understanding, and feeling are simultaneously active. It both grounds us in the immediacy of experience and carries us beyond it, producing a sense of transcendence that is inseparable from the experience itself.

This dual character—of simplicity and transcendence—poses a challenge for any attempt to reduce beauty to either subjective experience or objective structure alone. If beauty were merely a projection of subjective states, it would be difficult to explain its apparent object-directedness and its resistance to arbitrary variation. If, on the other hand, it were simply a set of structural features, it would be equally difficult to account for its capacity to evoke responses that exceed those features. The temptation to locate beauty exclusively on one side or the other—to treat it as either purely subjective or purely objective—thus risks overlooking the very phenomenon that requires explanation.

This tension is also evident in the limitations of more formal accounts of beauty. While the Thomistic framework, for example, provides a powerful articulation of the relation between proportion, integrity, and clarity, it may give the impression that beauty can be fully specified in terms of such features. Yet, as has often been observed, there are experiences of beauty that do not conform neatly to these criteria—instances in which what moves us is not harmony or symmetry but something more elusive, perhaps even arising in contexts of apparent disorder or incompleteness. To confine beauty to a fixed set of structural properties is therefore to risk excluding precisely those cases that most vividly display its power.

The point is not that structural features are irrelevant but that they do not exhaust the phenomenon. Beauty appears to arise in and through structure, yet it is not reducible to structure. It is encountered as something both immanent and, in some sense, transcendent: rooted in the organization of

14. Schindler, *Love and the Postmodern Predicament*, 40.

the object, yet not confined to it. This is why attempts to define beauty in purely formal terms often leave a residue—a sense that something essential has been omitted. The simplicity of beauty, in this respect, is inseparable from its resistance to complete analysis.

If this is so, then the universality of beauty, and its presence across a wide range of domains, need not be taken to imply that it is a vague or indeterminate concept. On the contrary, its ability to appear in diverse forms while retaining a recognizable character may be understood as a consequence of this very simplicity. Because beauty is not tied to a single set of features, it can be realized in multiple ways. Yet because it is experienced as a unified phenomenon, these diverse instances are not merely unrelated uses of a word but manifestations of a common underlying reality.

The Surfeit of an Experience of Beauty

The difficulty of reducing beauty to its constituent features is further illuminated by the observation that aesthetic experience often involves a sense of *excess*. John Milbank expresses this point with particular clarity in his discussion of beauty, noting that it arises where the attraction exercised by a formed reality escapes full analysis.[15] What is encountered in such cases is not merely a set of identifiable properties but a quality that exceeds any attempt to capture it in descriptive or conceptual terms.

This excess can be understood in two related ways. First, there is the sense in which no abstract account of beauty can replace the concrete experience of it. To describe the structural features of an object, even exhaustively, is not to reproduce the experience of its beauty. There remains a gap between description and encounter, one that cannot be bridged by the accumulation of further detail. The experience itself has a character that is irreducible to the elements from which it arises.

Second, there is the sense in which even the presence of the object does not fully satisfy the experience it elicits. Beauty, in this respect, appears to generate a kind of surplus: a response that is not exhausted by its occasion. One may continue to contemplate a beautiful object without the experience reaching a point of completion; indeed, it may deepen rather than diminish with continued engagement. This suggests that beauty is not simply a property that can be possessed in a finite and determinate way, but something that opens onto a horizon that extends beyond the object itself.

This dual aspect gives rise to what may be described as a paradox. Beauty is encountered *in* the object. It is not independent of the structure

15. Milbank, "Beauty and the Soul," 1.

that occasions it, yet it also appears to point beyond the object, to evoke a sense of something that is not fully contained within it. It is, in this sense, both immanent and transcendent: grounded in the world, yet not confined to it. To express this in more formal terms, one might say that beauty exhibits a structure that resists closure: it cannot be fully captured within the limits of any given instance.

From the standpoint of a strictly analytic framework, this may appear problematic, even contradictory. To say that something is both of the object and not of it seems to invite the charge of incoherence. Yet this reaction may reflect the imposition of a conceptual scheme that is not well-suited to the phenomenon in question. The fact that beauty resists straightforward categorization may be less a sign of confusion than an indication that it operates at the limits of our usual modes of analysis.

For some, this resistance will be taken as evidence that beauty is fundamentally subjective or emotive—a projection of human response onto an otherwise neutral world. For others, however, it may be taken to indicate that aesthetic experience discloses a dimension of reality that cannot be fully articulated in purely structural or causal terms. The excess of beauty, on this view, is not an illusion to be explained away but a feature that calls for interpretation.

If the latter perspective is adopted, then the presence of beauty within the structures described by mathematical physics takes on a particular significance. If beauty exhibits universality, resists reduction to formal criteria, and carries a transcendental dimension that points beyond any given instance, then its appearance within the most fundamental descriptions of the physical world is not a trivial matter. It suggests that the structures uncovered by science are not merely intelligible but are capable of eliciting a form of response that exceeds their formal articulation.

This brings us back to the broader framework within which this chapter is situated. If aesthetic properties are necessary constituents of structure, and if beauty represents a particularly intense or elevated realization of such properties, then the presence of beauty at the foundations of the physical world invites reflection on the nature of those foundations. In particular, if beauty is associated with the disclosure of meaning, value, and even a form of transcendence, then its ubiquity may be taken to indicate that these dimensions are not accidental features of reality but belong to its basic constitution.

Within the theological tradition, this line of thought has often been developed in terms of the relation between aesthetic, moral, and ontological apprehension. As Catherine Pickstock notes, these modes of engagement have been understood as rooted in a more fundamental relation to the divine. Beauty, in this context, is not an isolated phenomenon but one of the ways in which the structure of reality is disclosed as meaningful and, ultimately, as expressive of a deeper source.

The convergence of these considerations—universality, structural grounding, irreducibility, and transcendence—suggests that beauty occupies a distinctive position within our attempt to understand the world. It is present in the objects we encounter, in the theories we construct, and in the experiences through which we engage with both. It resists reduction to any one of these domains, yet connects them in a way that is difficult to ignore. For this reason, beauty may serve as a bridge between the physical, the metaphysical, and the theological, providing a point at which questions about structure, meaning, and intentionality converge.

Concluding Section: Beautiful Structures

The convergence of these considerations—beauty's universality, its grounding in structure, its resistance to reduction, and its persistent transcendence of any particular instance—suggests that it occupies a distinctive and irreducible position within our attempt to understand the nature of reality. It is not merely one feature among others nor an optional embellishment added by human sensibility to an otherwise neutral world. Rather, it appears as a condition under which the world becomes intelligible at all: a mode in which structure is not only apprehended but recognized as meaningful. In this sense, beauty does not follow from our understanding of the world; it belongs to the very possibility of there being a world that can be understood.

If this is so, then the argument developed across this chapter extends beyond the classification of aesthetic experience and into the deeper question of what it is for anything to exist as a structured reality. For if all structure is necessarily aesthetic, and if aesthetic properties are intrinsically normative, admitting of better and worse realizations, of proportion and distortion, of integrity and failure, then normativity is not an addition to the world imposed by human judgment but a feature of the world's own constitution. The aesthetic, in this sense, becomes a primary site at which the unity of fact and value is disclosed: the point at which the structure of things is already evaluable, already bearing within itself the marks of order, fittingness, and intelligible completion.

This has direct consequences for the broader thesis of this work. The argument has not been that beauty provides an isolated "pointer" to something beyond the world nor that it functions as a probabilistic indicator within an evidential framework. Rather, the claim has been that the very conditions under which anything can appear as structured—conditions without which neither science, nor philosophy, nor any form of discourse would be possible—are already suffused with aesthetic character. To recognize this is not to add a theological layer to an otherwise self-sufficient natural order, but to see that what we call the natural order is itself constituted in such a way as to invite, and indeed require, interpretation in terms that exceed purely causal description.

It is at this point that the analogy with artifact, which has been present throughout, reaches its full significance. In the case of human artifacts, the recognition of purposive form is not the result of an inference from neutral data but is given in the perception of the object itself. We encounter the work as expressive of intention; the presence of the personal is not deduced, but seen. If the structures of the natural world share, at a fundamental level, the same aesthetic conditions; if they exhibit forms that are not only intelligible but normatively ordered; and if they give rise to experiences of beauty that carry a sense of surplus and transcendence, then the distinction between artifact and nature begins to lose its absolute character. The possibility emerges that what is disclosed in the one case may, in a transformed but continuous way, be disclosed in the other.

The suggestion, then, is not that the world is like an artifact in some loose or metaphorical sense but that the conditions under which we recognize artifactual form—conditions rooted in aesthetic structure and normative evaluation—are already operative in our encounter with the world as such. The step to interpreting this in terms of intention, and ultimately of divine intentionality, is therefore not an arbitrary extension but a development internal to the logic of the experience. The language of divine artistry does not impose an external framework upon the world; it articulates what is already implicit in the way the world is given to us as structured, intelligible, and, at its limits, beautiful.

In this light, beauty can be understood as the visible—or more precisely, the perceptible—condition of intelligibility itself: the moment at which structure achieves a form of coherence that both satisfies and exceeds our capacity to comprehend it. It marks the point at which order is not only present but recognized as order, and in that recognition carries with it a surplus of meaning that cannot be contained within the structures that occasion it. This surplus is not an accidental by-product but an intrinsic feature of the aesthetic: the indication that what is disclosed is not exhausted by its manifestation.

To speak of transcendence in this context is therefore not to invoke a realm beyond the world but to acknowledge that the world, in its very intelligibility, points beyond any finite articulation of it. Beauty is the mode in which this pointing is most acutely felt. It is at once the grammar of structure and the sign of its excess, the condition under which things can be known and the reminder that they are never fully known. In this way, the aesthetic becomes the site at which the intelligibility of the world and its inexhaustibility are held together.

If the argument of this chapter is correct, then the presence of beauty across the domains of mathematical physics, natural form, and human creation is not a coincidence, nor a mere feature of human response, but a reflection of a deeper unity in the nature of reality. The structures that science uncovers are not only mathematically describable but aesthetically apprehensible; the forms we create are not only functional but expressive; the experiences we undergo are not only subjective but responsive to features that belong to the world itself. Taken together, these suggest that reality is not merely ordered, but ordered in such a way as to be recognizable as order: available to a form of understanding that is at once cognitive and evaluative.

At this point, the broader claim of the book comes into sharper focus. The intelligibility of the world—its susceptibility to mathematical description, its openness to scientific investigation, its capacity to sustain coherent discourse, is not a brute fact but a feature that stands in need of elucidation. If that intelligibility is, as I have argued, inseparable from aesthetic structure, and if aesthetic structure is intrinsically normative and, in its highest expressions, expressive of a surplus that points beyond itself, then the world presents itself not as a closed system of causal relations but as something more akin to an intelligible whole whose order is, in a significant sense, given.

To recognize this is not yet to complete a theological argument, but it is to locate the point at which such an argument becomes both possible and, perhaps, unavoidable. For if the world is structured in such a way that its intelligibility is inseparable from aesthetic normativity, and if that normativity carries within it the marks of purposiveness and expressive form, then the question of whether this structure is ultimately grounded in something analogous to mind or intention cannot be indefinitely deferred. The experience of beauty, in this context, is not a distraction from that question but one of the most direct ways in which it is encountered.

Beauty, then, is neither an ornament of understanding nor an optional addition to it. It is the point at which understanding recognizes itself as responding to a world that is already ordered, already meaningful, and already, in some sense, given in a form that invites recognition. In that recognition, the divide between fact and value, between the physical and the personal, begins to dissolve, and the possibility emerges of seeing the world—not as a collection of neutral structures but as an intelligible whole whose very being is bound up with the conditions under which it can be known, evaluated, and, ultimately, understood as the expression of a deeper source.

It is precisely this convergence that is captured, in compressed form, by the title of this work. To speak of *beautiful structures* is not merely to ascribe an occasional property to certain objects but to recognize that wherever structure is found, it is already open to aesthetic characterization, and that in its most revealing instances this character takes the form of beauty. The structures disclosed by mathematical physics, in particular, are not only formally precise and empirically powerful but are repeatedly encountered as exhibiting a form of elegance, unity, and inevitability that calls for aesthetic recognition. In this sense, the *aesthetics of mathematical physics* is not an external gloss upon scientific practice but an articulation of something internal to the very structures through which the physical world is made intelligible.

At the same time, to speak of *the art of creation* is to acknowledge that the language we ordinarily reserve for human making—language of form, intention, expression, and purposiveness—cannot be entirely confined to the domain of human artifact. The argument of this chapter has been that the conditions under which we recognize artifactual form are continuous, at a fundamental level, with the conditions under which we encounter the natural world. If beauty, as a mode of aesthetic disclosure, carries within it both normativity and a surplus that exceeds its immediate occasion, then the presence of beauty at the foundations of physical reality invites a mode of understanding in which the distinction between artifact and creation is no longer absolute but gathered into a deeper unity.

The title thus gathers together the central trajectory of the argument: that the intelligibility of the world, as disclosed with particular clarity in mathematical physics, is inseparable from its aesthetic character, and that this character is not merely descriptive but expressive. The structures we encounter are not only there to be analyzed but are given in such a way as to be recognized—as coherent, as ordered, and, in the fullest sense, as beautiful. Where such recognition takes place, the language of art—of making, of intention, of creation—arises not as an imposition upon the world but as the most adequate way of articulating what is already implicit within it.

To speak, then, of *Beautiful Structures: The Aesthetics of Mathematical Physics and the Art of Creation* is to name, in a single phrase, the claim that has been unfolding throughout this work: that reality is not merely structured, but structured in such a way that its order is apprehensible as meaningful, its intelligibility as beautiful, and its presence as, in some profound sense, the mark of creation.

PART 2

Beauty and Natural Theology

It is perhaps surprising given these reflections that, as Philip Tallon notes in a recent paper, beauty "has rarely been the major theme of any major modern theologian's or philosopher's thought."[16] De Cruz and de Smedt support this contention. They suggest that "the argument from beauty enjoys relatively little philosophical interest and discussion."[17] They go on to note that no major anthology of natural theological arguments incorporates a substantial argument from beauty, let alone an argument from aesthetics to God from the natural sciences specifically. The formal literature is sparse.[18] Tallon offers a telling example of this in his paper. He writes,

> The most vivid instance is "The argument from aesthetic experience" in Kreeft and Tacelli's "Twenty Arguments for the Existence of God." Though nineteen other arguments are given multi-page treatments, or at least a full paragraph, the argument from beauty is only three lines long: "There is the music of J. S. Bach. Therefore, there must be a God. You either see this or you do not."[19]

There is of course an amusing brevity to this, but it echoes a recurring theme in how beauty is frequently treated in the natural theology literature. There is nevertheless a certain truth to Tallon's argument. Understanding what makes this so, though, yields more of interest than what the compressed argument suggests is to be understood.

What is an "argument from beauty" for God? The phrase "argument from beauty" is described by de Cruz as indicating "a family of arguments for the existence of God that takes the beauty of the natural world or works of art (e.g., Bach's cantatas) as evidence for the existence of god."[20]

The argument from beauty is different from traditional classical arguments for the existence of God which begin from some general fact about the world and deductively concludes God's existence. Intense beauty of both humanly derived objects and others often causes a psychological and

16. Tallon, "Mozart Argument," 321.

17. De Cruz and de Smedt, *Natural History*, 133.

18. A good example of what issues the place of beauty in theology is commonly taken to concern are discussed in Viladesau, *Theological Aesthetics*.

19. Tallon, "Mozart Argument," 322.

20. De Cruz and de Smedt, *Natural History*, 131.

intellectual movement beyond the beautiful thing: a feeling that the object itself cannot fully contain what it provokes.

The philosopher Anthony O'Hear explicitly acknowledges the power of beauty to apparently point beyond itself:

> The religious impulse might be a response to a natural object or work of art as beautiful: in which we see a thing not just as a . . . "fragment of nature," but something which mediates between our own longing for perfection and some other world in which that perfection is realised.[21]

O'Hear's position is echoed by Bruno Forte: "'Beauty' is an event; beauty happens when the Whole offers itself in the fragment, and when this self-giving transcends infinite distance."[22]

As de Cruz acknowledges,

> unlike other arguments for the existence of God, which move from clearly stated premises to conclusions, the argument from beauty has an ineluctable phenomenological component, the experience of beauty and awe and its connection to religious experience. The connection between beauty and God's existence is less directly obvious, even if it is convincing when one has an aesthetic experience.[23]

She captures something of this sense when she writes,

> The books and music (and nature) in which we believe beauty is located do not contain the object of the desire. Rather it is through them that we glimpse the unreachable object we long for. The sense of the sublime that is present in our experiences of great works of art or natural beauty can be regarded as a manifestation of *Sehnsucht* . . . a sense of being overpowered by one's aesthetic experience and one's longing (and inability) to be part of what one is experiencing.[24]

Even those who do deny the existence of God often admit the curious effect of beauty to provoke transcendental thoughts and feelings. Tallon writes that

> even among atheists the argument from beauty has a persistent persuasive power that remains when more traditional

21. O'Hear, "Science and Religion," 512.
22. Forte, *Portal of Beauty*, vii
23. De Cruz and de Smedt, *Natural History*, 132.
24. De Cruz and de Smedt, *Natural History*, 134.

> arguments for God's existence falter. . . . Akin to theists, who do not quite know what to do with beauty, yet remain fascinated by its . . . power, atheists often seem haunted by the theistic argument from beauty.[25]

Paul Draper, an atheist, even more strongly suggests that beauty is *evidence* for God. Nevertheless, he thinks that on balance, "arguments from evil against theism are much more powerful than the argument for beauty in favour of theism."[26]

Many atheist scientists, like Neil deGrasse Tyson—an astrophysicist, the director of Hayden Planetarium, and a popular science writer—recognize a transcending existential urge provoked by their disciplines. Some, like him, have even tried to offer a "secular spirituality" to provide some outlet for these feelings.[27]

Other nontheists have labored to produce what they call "religious naturalism," the idea that it is possible to develop religious feelings and ideas about the natural without recourse to any concept of a divine "ground-of-being," the world itself being sufficient to supply that.[28] The need to make sense of the urge is perhaps evidence of being in its "natural" forms calling forth a desire to understand its shapes, origin, and our own nature as participants in the world.

The uncovering of the scale, intricacy, and profound intelligibility of the natural world, particularly in the field of physics, as I have discussed, produces for even nontheistic scientists expressions of "awe in the face of nature"[29] that normally incorporate reference to aesthetic qualities, and often profoundly moving existential reflections on the nature of man, the splendidness of the cosmos, and the place of *Homo sapiens* within it, whether atheist or theist.

The physicist Stephen Barr, considering the property of "symmetry" in art and physics, opines,

25. Tallon, "Mozart Argument," 323.

26. Draper, "Seeking but Not Believing," 204.

27. Keas, *Unbelievable*, 102–6, contains an interesting discussion of Tyson and others under the heading "The Copernican Demotion Translates into Spiritual Demotion."

28. Stone, *Religious Naturalism Today*, provides a description and defense of a religious naturalist position.

29. Ecklund, *Science vs. Religion*, 58, provides a number of interesting findings. Among scientists who described themselves as "spiritually" engaged, 22 percent identified as atheists, 27 percent as agnostic, and 51 percent of the remainder identified as theists. It is therefore at least clear that what the sciences uncover often prompts existential responses in investigators that transcend the immediate properties of the objects they seek to explain, however one chooses to interpret this fact.

> If symmetry is found in works of art of every sort, and is an important element in what is beautiful, and if laws of nature are based on symmetries so sophisticated . . . that while we may study them with mathematics, they lie far above our powers to appreciate on an intuitive level, does that do not suggest the mind of a "divine" artist at work?[30]

What Barr is trying to get at is the transcendent feeling that can be provoked by the beauty of the intelligible structures of the particle world which present themselves as bearing all the features of intentional construction. The intuition of scientific theists like Barr is that an aesthetic experience of this kind is naturally revelatory of the presence of a divine author.

Saint Augustine thought the perception of beauty was of unique significance in our experiences of the natural world about us in generating desire for the Creator.[31] Augustine thought of creation as a "great book that is open before us."[32] Of the beauty of nature that moves us towards the divine he has this to say:

> God did not make letters of ink for you to recognize him in; he set before your eyes all these things he has made. Why look for a louder voice? Heaven and earth cries out to you, "God has made me."[33]

There is in Augustine's description of things of natural beauty "crying out" a supremely important observation: that natural events can express personal things. "To cry out" can only be done by objects whose structures have been deliberately shaped and created to elicit feelings and thoughts that something of value is being expressed. We should be careful how we understand what is meant by "structures." What Augustine means is not that every object of nature is carefully crafted by God's hand but that it is in the very existence of contingent natural structures at all that they "cry out" their creation.[34]

30. Barr, *Modern Physics and Ancient Faith*, 104. It is perhaps necessary to discharge another potential objection: beautiful objects *can* be accidentally produced in nature, but the possibility of such accidents is *still* dependent on their being underlying aesthetic features in natural structures, the coming together of which by chance can produce beauty. These would have to *preexist* any such occurrence, so inference to the divine would not be affected by this argument.

31. Ramirez, "Augustine's Proof for God's Existence," 122.

32. Travis, "Glorious Resonance," 25.

33. Augustine of Hippo, *Sermons* 68.6 (Hill, 225–26).

34. See also ch. 1 ("Divine Beauty: Augustine") and ch. 2 ("The Word of Beauty: Thomas Aquinas") in Forte, *Portal of Beauty*.

When we encounter great beauty in events of nature, the effect of a personal hand behind them is intensified. Physicists like Barr draw theistic conclusions from the intricacy, scale, and aesthetic structures of physics because they are moved not only by the intelligible coherence of those structures but also by the thought they could have been "ugly." That they are not at least supports, if it does not prove, that there lies behind this feature of the natural an intention that it should be so.[35]

A quote from the American Quaker Rufus Jones sums up the answer with a more appealing simplicity:

> Beauty implies behind things a Spirit that enjoys beauty for its own sake and floods the world everywhere with it. Wherever it can break through, it does break through and our joy in it shows that we are in some sense kindred to the giver and revealer of it.[36]

The theme that beauty in nature elicits not only powerful feelings that are experienced as being provoked by the beauty of natural objects themselves but also as pointing to an *author* who embellished them with beauty, has also been championed by the theologian and analytical philosopher Alvin Plantinga. He suggests that the beautiful and awe-inspiring in nature do not serve as a *premise* of a deductive argument but rather are non-inferentially "occasioned by the circumstances; they are not conclusions from them. The heavens declare the glory of god and the skies proclaim the work of his hands: but not by way of serving as premises for an argument."[37]

The Common Language of Beauty, Nature, Artifacts, and Transcendence

A marvelous example of a human artwork that is at one and the same time dependent on both "natural" features of the geographical environment and human artistry to try and attain and communicate a sense of transcendence

35. The argument that microphysics (or cosmology), or the structure of the cell, for example, shows evidence of beauty will have to contend with the objection that the natural world also shows evidence of "ugliness." The argument will require a "theodicy of ugliness." In other words, it will have to answer the question why a good god, appreciating beauty, allowed for ugliness, or created "ugliness" or allowed for its possibility elsewhere in nature. I do not underestimate the importance of this observation, but my purpose here is to examine the presence of beauty in physics, not to account for ugliness.

36. Stevenson, *Open to New Light*, 133.

37. Plantinga, *Warranted Christian Belief*, 175.

is the prehistoric cave paintings of Lascaux, dated to fourteen thousand years ago.[38]

Photograph of Lascaux painting from the Lascaux Caves, France[39]

Although it can never be entirely clear whether the cave art was constructed for pleasure, an animistic attempt to manipulate nature, an expression of the regard for the animals on which they relied so heavily for life, an attempt to assuage or contact spiritual forces, or all of these things, the mystical transcendent quality of the illustrations is immediately apparent to us. The force of the paintings to most observers seems like a poignant attempt to transcend the mundanity of the everyday, whatever their ultimate purpose.

It is clear that the picture is constructed with a careful and intentional haunting beauty; it is intended to manifest the hidden in the forms and colors of the cave itself. The aesthetic appeal derives from the way the natural colors, lighting, and contours of the cave interior have been brought together with the features and movements of the animals.

It is a beauty that transcends the materiality of its composition, the artists' intention, and even the creatures themselves. The creation of the painting below the surface of the earth has an intuitively gripping sense of

38. Wagner and Briggs, *Penultimate Curiosity*, 23.

39. Lascaux painting, courtesy of the EU, https://commons.wikimedia.org/wiki/File:Lascaux_painting.jpg, licensed under CC BY-SA 3.0, https://creativecommons.org/licenses/by-sa/3.0/deed.en.

"going below" to uncover the inner mystery of the surface. In other words, a physical embodiment of the sense of something behind or beyond the everyday.

Spiritual and religious art of all kinds and in all cultures is often the attempt to make manifest the meaningful and the ultimate using the expressive forms and structures of the natural world. In the history of the Judeo-Christian tradition art, music, and the written word have of course been associated with attempts to convey something of the divine *through* beautiful composition and sometimes by analogical portrayal of the divine in human shape. In one thread of a conventional Christian tradition, of course, the Son of God himself is revealed in corporeal and natural form.[40]

Whatever view one wishes to take of theistic or atheist conclusions about natural sciences, the aesthetics of nature, and human creations, it is undoubtedly the case that interspersed among the evaluative language used to describe or judge structures in any of them are references to mutual structural features. A delightful example comes from David Stipp's book on Leonard Euler, the mathematician.[41] Stipp's book is titled *A Most Elegant Equation: Euler's Formula and the Beauty of Mathematics.*

The interior front cover of the book is awash with references to beauty, art, and structure. Euler is there described as "the Mozart of mathematics," his famous equation as "a conceptual diamond of unsurpassed beauty." "Few," the paragraph tells us, "have been left unmoved by it." The mathematician Keith Devlin compared it to "a Shakespearean sonnet that captures the very essence of love."

It is significant that the wash of metaphors relates to one equation, with one author, whose mathematical artifact provokes an aesthetic response that embraces the human, the physical, and the abstract all contained in one small structure. These metaphors express a sense of awe that the brief squiggle of shapes making up the equation could have such profound and far-reaching effects across so much of our experience of the world, yet could issue from the interior of one creative and analytic mind.

Incorporated into that sense of awe is a wonder that squiggles, nature, beauty, and the descriptive and predictive power that issues from Euler's equation should be possible at all. The reaching for metaphor is an attempt to convey a thought, whether implicit or explicit, that the significance of Euler's equation lies not merely in what it accomplishes but in the fact that it exists at all. That a small arrangement of symbols should unite beauty,

40. Harries, *Art and the Beauty of God.* The former bishop of Oxford considers the relation of beauty in art in the light of the unique revelation of Christ.

41. Stipp, *Most Elegant Equation.*

intelligibility, and such remarkable descriptive power over the physical world can itself provoke a sense of wonder. The intuition behind the metaphors is that the conjunction of mathematics, nature, and aesthetic value is not simply a mundane fact to be observed, but something that, by its very character, seems to call for a deeper explanation than its mere presence.

A thought, perhaps, that only something transcending the mathematics, syntax, symbol, and interpretation, and its structural applications both physical and otherwise, would be sufficient to meet the power of appreciation that it evokes: something transcendent.

Divine Origins: The Unity of the Natural Sciences, Mathematical Structures, Art, and Beauty

An interesting if perhaps unintentional revelation of the assumed closeness of the divine author, the natural sciences, mathematics, artistic creation, and beauty in the theist tradition is uncovered if one searches for "the argument from beauty" in the online *Oxford Handbook of Natural Theology*.[42] It takes one directly to a section titled "Aesthetics in the Arts." Frank Burch Brown begins his contribution to that section in the following way,

> The design of the world would seem less wondrous, less divine, if it were devoid of beauty. When Augustine and medieval Christians under his influence referred to the biblical book of Wisdom, and its assertion that God has made all things according to the measure, number, and weight (Wisdom 11:20), they partly had creation's beauty in mind. In their thinking, number and mathematics were not divorced from the other qualities of beauty such as its harmony and luminosity.[43]

The use of words like "design," the reference to number and mathematics, "harmony," and "luminosity" are all descriptions, in one way or another, that can only make sense if their objects have intrinsic structures, mathematically expressible, whether or not they refer to human creations or the structures of natural and abstract worlds.

This invites the question: From whence do the standards of beauty derive if we do find mathematical structures of nature that merit the accolade of beauty? If they do, this invites the further inquiry: Why should the beauty in our creations share aesthetic judgments with those of natural objects? These are questions about the origin of aesthetic normativity.

42. Brown, "Aesthetics and the Arts."

43. Brown, "Aesthetics and the Arts," 611.

In previous chapters I have been concerned to argue that anything which comes into existence must have an intelligible structure and that anything that has a structure must, pari passu, have an aesthetic characterization. I have also tried to present a persuasive case that the necessity for structure brings with it normative standards. Normative standards must be, on this account, necessarily present when anything of physical (or abstract) substance is present to us.

This is, I have maintained, because anything that can be judged according to some standard must make appeal to something which transcends that which is being judged: to judge something as "beautiful" or "ugly" is to refer to an idea of what is to count as beautiful and weigh the thing being judged against that standard.

In the case of the sciences, to judge something as "evidence" for a hypothesis is to have some prior standard of why something is evidence and something else not. "Induction," "deduction," and "abduction" are all epistemological methods which rely on the fact that they stand in certain kinds of epistemological relation to the world and yield under certain conditions what is to count as "knowledge" of that world. Each one of them rests on the assumption that they provide standards of cognitive attainment.

Mathematical features of structures provide another normative relation; if one accepts that mathematics cannot be reduced to any description of any given set of physical properties, then the features of structures in virtue of which their mathematical description is enabled must not themselves be "physical" in the way required by the physicalist naturalist.

I adumbrated the issue in my chapter on aesthetics in physics like this: Since mathematical entities, however conceived, are held to be acausal, outside space-time, and nonmental, how physics can be claimed at one and the same time as evidence for the assertion that the universe has no basic ontological space for nonphysical domains is at best problematic and at worst a contradiction deriving from the argument's own empirical base. Susan Schneider, whom I mentioned in an earlier chapter has advanced this case in a paper called "Does the Mathematical Nature of Physics Undermine Physicalism?,"[44] calls this the "Problem of the Base." It provides prima facie difficulties for an ontological naturalism that insists on a reductive physicalist base on which everything must rest.

What, one might inquire, could provide the normative standards if these are *requirements* for the existence of structures *and* the standards by which they, at *one and the same time*, can be judged? Could the provision of such standards be provided by the physical universe itself? Since the

44. Schneider, "Mathematical Nature of Physics."

coherence of anything in the structured physical universe depends on the provision of normativity for its emergence, this is not a logical possibility. Could it emerge from physical structures, perhaps of sufficient complexity? Not only can the last argument be reprised to deny this possibility, but since complexity itself is a structured outcome of a series of events, this would be putting the explanatory cart before the horse.

"Normativity" implies the possibility of judgment, a comparison between what could be and what is. To this extent, it also implicates the presence of a rational mind able to grasp the abstract standard, apply it to the object or event in question, and reach a judgment as to what degree the object being judged attains the standard and in what ways it has succeeded or failed. Anything that can engage in logical thought must have features of rational mindedness. That, after all, is a precondition on making any sense of what is meant by "standards": the capacity to make comparative judgments.

Likewise, and for similar reasons, structures at the base of the physical world which manifest rational beauty reveal our innate tendency to suppose that items of great beauty cannot be accidentally so but must portray intention. After all, something beautiful *could* have been ugly. A good explanation of this is that behind what is beautiful stands a mind that can distinguish between the two.

If, by a pari passu doubling, firstly, the fundamental structures on which all matter depend in one way or another are characterized by the aesthetic, and secondly, the empirical base of the physical world turns out to be beautiful and theoretically useful, then this is at least supports a rational reaction that what is being confronted is a work of imaginative intelligence.

Perhaps in these circumstances, it is possible to consider the relation between man and artifact and God and natural world as *sharing* the same characteristics that press for a teleological and meaningful comprehension of what the created object expresses. The shared meaning between understanding human authorship and creativity and that which is elicited by the phrase "proclaim the work of his hands" is so tightly shared that its interpretation as metaphor or analogy does not seem adequate to the content which is intended. The meaning hinges on the term "create," and that hinge itself suggests a participation in authorship.

Divine Creation, Human Creations, and Beauty

Finally, I want to end this section by making clear an important distinction between being moved by the beauty of a human artist's creation and being moved by God's creation in nature. In the former the object is entirely within our experience of what we count as our day-to-day world. Our experience of the author *in* the beauty is of something that is wholly us, *Homo sapiens*, another human person.

When we have a transcending experience[45] of beauty that points to God, however, we have to be careful how we understand the "externality" of "nature" and what the transcendentality of the experience involves. Because we ourselves are a part of God's creation, *we are at one and the same time an external "object" of God's creation as well as its observer.* The appropriate word for this is perhaps "participation": we are at one and the same time individual creations but *participants* in God's world.[46] As such we are privileged to share in part of God's nature and abilities. We are made, as the traditional phrase has it, *imago Dei.*

The idea of a human creator is of an artist who intentionally manipulates the material *with which she is presented* to create an object. These materials are all around her in the natural elements of the world and those artificially constructed from those elements. She also uses principles of structure, the geometry of perception, the artistic fashions of her time and place, and so on to create what it is she wants to be seen by the observer. It is a mistake that easily follows from this to take God as someone who manipulates raw material into designed forms.

God does not *replicate* the human creator. The theist God does not create the world from preexisting materials like Plato's master craftsman of the *Timaeus*, the Demiurge, but creates ab initio. It is important to be clear what this means, whatever other problems it involves. It is a claim that *whatever* exists, God brought it into existence. This is hardly a new thought. The ante-Nicene patristic father Irenaeus (AD 130–202), among others, made it quite explicit:

> Men and women cannot make anything out of nothing, only out of matter that exists; God, however, is far superior to humankind inasmuch as he himself invented the matter of his work, since previously it did not exist.[47]

45. I have concentrated only on "beauty" and transcendence. This is not to say that there are not many other features of world and existence that can provoke such feelings.

46. The idea of "participation" in God is developed in rich detail by Andrew Davison. It would do the book an injustice to attempt a full summary here, but one distinction Davison draws is between being "part of" something and having a "part in" it. The latter carries a stronger sense of participatory inclusion and can accommodate the idea of being fully involved while nevertheless not constituting the whole, perhaps analogous to the way in which a member of a team is fully part of the team while not being identical with it. See Davison, *Participation in God*, particularly ch. 6 ("Characterizing Participation") and ch. 7 ("Analogy: Participation in Being and Language").

47. Irenaeus, *Against Heresies* 2.10.4 (Unger and Dillon, 37).

In a discussion titled "What Does Physical Cosmology Say About Creation from Nothing?," Adam Hincks makes the point this way: "God is not just a Platonic demiurge that informs preexisting matter; rather he is the creator of matter and the author of the very existence of possibility itself."[48]

The idea of *creatio ex nihilo* in which matter and space-time present themselves as part of a coherent and comprehensible universe issuing from the hand of divinity, has long featured as part of a unifying theist explanation of the cosmos, as the quote from Irenaeus illustrates. It is an explanation of the very possibility of "being" and its various aesthetic structures of existence. The universe, on the theist account, presents itself to us as an accessible and meaningful production of a ultimate divine mind in which its presentation to us is of a unified and contingent body open to coherent articulation and investigation. The intent that it should be so is displayed in the cohering beauty that the fundamental structures of the universe broadcast as part of a signature of divine authorship.

48. Hincks, "What Does Physical Cosmology Say," 328.

Chapter 5

A Metaphysically Designed World

Teleology and the Personal in Arguments for God's Existence

SUMMARY

In this chapter I aim to begin to defend and apply the notion that teleology and divine personhood lie at the uniting heart of an ultimate conception of the world as "God's artifact." Because this idea invokes the concept of intentional design it can, wrongly construed, appear as an explanatory hypothesis in competition with the assumptions of scientific methodology.

"Intentional" design and the personal response to it, I argue, is, or should be, elicited by the *entirety* of creation. This includes the *presence* of physical structures, mathematically intelligible and aesthetically pleasing. The emphasis on the word "presence" is to forgo a conflation between the metaphysical necessity that any existing thing must have *some* structure, and scientific investigation which uncovers the *kinds* of contingent structures that do exist and their causal functioning.

Because it is possible for events, particularly in mathematical physics, to be understood as objective events *and* as manifesting beauty, it is plausible to see them as ultimately an expression of divine teleological purpose. We respond to them as we might respond to any beautiful artifact constructed by human persons: with a resonating sense of the intention that is expressed through the beauty of the construction.[1]

1. I should reiterate that we also need a theodicy of ugliness. There is much that is ugly.

OVERVIEW OF THE CHAPTER

This chapter turns on a central pivot: the thought that the recognition of another as a person is not dependent on first establishing unassailable criteria of personhood. The artifacts of human life similarly manifest the agency of the personal with an immediacy that confirms their teleology more persuasively than any proof could do.[2] As Ridley argued, "The artist must be seen as present in his work, much as a person must be seen as present in his behaviour, rather than as separate from it, behind it, or, above all, logically independent of it."[3]

It is in the light of the humanity the created object displays that we can distinguish teleological artifacts from those of inanimate provenance. None of these epistemological observations reduce "personhood" or artifacts to a secondary ontological status. As I have argued the position that it does is the consequence of inflating methodological naturalism to an ontological conclusion.

The distinction we make between our artifacts and those that are the outcome of nonteleological processes is a distinction we make between objects from our vantage point. Seen in divine aspect, however, we recognize that the aesthetics of the physical world, our own artifacts, and ourselves can resonate with the divine creative intent that makes them possible. In this perceptual mode we see the entirety of being as the outcome of divine intention: ultimately God's artifacts.

Whether someone can be bought to this experience by deductive rational proof alone is moot. Beauty in objects, for example, as I have previously discussed, can move one beyond the beauty of the thing itself to the sense of something it cannot contain: the presence of the transcendental authorship which gave it life—a personal agent.

The rationality of that reaction is perhaps more persuasive of the presence of personhood than deduction alone. The sense of another person is not a matter for objective decision-making but the *outcome* of an engagement with them and their world. Further, our own sense of being a person is dependent on our participating in the world of other persons. Our

2. There will be "gray" areas, of course, where the provenance of an artifact is uncertain. The reader will, however, recall my distinction between "distinction" and "dichotomy" and the discussion in chapter 4 about concepts. In brief summation: from the observation that we are sometimes uncertain as to whether something is a "chair" or another item of furniture, it does not follow that we are not able to use and identify chairs without difficulty.

3. Ridley, "Expression in Art," 218.

participation in God's artifact is an echo of that relation we have with other persons.

The Cartesian picture of a consciousness separated from the world and others is a *philosophical* account of the relation between mind and body, not an inevitable accompaniment of the nature of consciousness itself. The experience of *that* separation is consequent upon recognizing that there are others to feel separate from. This is a common *psychological* experience but not a proper conceptual basis for the concept of the person (I discuss why this is further on).

While an experience of God's person in nature is not that of an encounter with another human person, like that encounter it is manifested in and through the shape, activity, and presence of objects as physical structures, and of course ourselves as corporeal living entities.[4] The experience of beauty in mathematical structures at the world's base enhances and sharpens that sense, involving the thought that beauty is intended to manifest something valuable and good.

In pursuit of enhancing the plausibility of this claim, I address a number of potential objections to it. Perhaps one of the most important of these is the idea that our tendency to perceive personal agency in the forces and structures of nature, whether in the form of gods, monotheism, spiritual forces, and so on, has nothing to do with the reality of *any* of these things but is better approached via evolutionary considerations. Once the evolutionary heritage and underlying emotive functions of "spiritual" beliefs are properly understood, it is supposed, theistic urgings can be explained without reference to God. I contest this view.

I move on to show that many other arguments for God make necessary reference to teleology and the personal at some point in their exposition. The "fine-tuning" version of the cosmological argument is a substantial

4. Rea, *Hiddenness of God*, presents an argument for the idea that the physical functions as the vehicle for the expression of mindedness. This is not equivalent to the claim that features of mindedness are identical with, reducible to, dependent upon, or merely properties of neurological tissue.

Hacker and Bennett, in *Philosophical Foundations of Neuroscience*, offer a sustained critique of the manner in which the brain–mind relation is understood within much contemporary neuroscience and philosophy of mind, identifying a number of conceptual confusions. In broad outline, their criticism is that much modern neuroscience and philosophy of mind simply recoils from one side of the Cartesian divide—mind and consciousness—to the other—namely, the physical. The figures they discuss include eminent thinkers such as Colin Blakemore, Antonio Damasio, Michael Gazzaniga, and Francis Crick. The book generated extensive discussion in both philosophical and neuroscientific circles, attracting strong praise as well as severe criticism. An accessible exchange between the authors, John Searle, and Daniel Dennett may be found in Robinson, *Neuroscience and Philosophy*.

example. The argument contends that there are features of the universe whose exact coappearance at the emergence of the universe can only be explained by divine intention.

I also briefly examine the ontology of morality. My point here is not to engage with the extensive literature in moral philosophy but to show that the idea of a personal god sensitive to the guidance of personal relationships and obligations can provide for the justificatory power of appeals to a personal law giver.

Very lastly, I consider the recent argument by Michael C. Rea who has written, "The only way we manage to experience the presence of other minds within the material world is by way of cognitively impacted experiences whose stimuli are natural, physical phenomena."[5]

Rea contests the long-standing idea that this yields an insurmountable logical and empirical gap between "person" and the "physical" as I have been concerned to do. His quote ends with the conclusion "that reflection on other minds points to the conclusion that we would experience God's presence in much the same way as we experience the presence of created persons."[6]

I discuss his argument and go on to show how my conception of teleology in artifacts and nature, particularly that of beauty, provide ways in which that personal encountering can take place.

PART 1

The Basic Concept of the "Person"

The idea that the aesthetic structures of sub-particle physics cause us to have the same affective responses as our experiences of the artifacts of human agency and that they express divine intention will be a source of considerable skepticism for some. For the atheist and the naturalist who disavows that the impersonal could be penetrated by the personal in the way this suggests, it will be confirmation of the essential "irrationality" of religious belief. The existence or otherwise of God, it will be mooted, is not a matter for decision on the basis of nonrationally explicative experience.

John Cottingham captures the essence of philosophical critiques of religious belief as turning on this point. "They construe," he says,

5. Rea, "Divine Presence," 132.

6. Rea, "Divine Presence," 132.

> acceptance of a religious outlook . . . mainly in terms of intellectual assent to a set of theses or doctrines. And what the critic tries to do . . . is extract the juice, as it were—the relevant set of theses—from what is taken to be a largely irrelevant background "pulp" of emotive, poetic, narrative, and symbolic elements.[7]

There is an analogy between philosophical attempts to provide deductive proofs for the existence of minds and persons other than oneself. Proofs of God's existence can be regarded as needing a similar justification to the perception of the hand of the divine personal in "nature" as "proofs" of other minds have in our relations to other persons: requiring deductive proof before one can be sure the other is indeed a person.

If, however, the conviction of encountering the divine in the world about us (and in ourselves as natural creatures) is akin to the interpersonal engagements we have with others, then "intellectual assent" to a set of rationally defensible "theses or doctrines," while not beside the point, is not the basis from which mutual recognition begins or, for that matter, from which self-identification as a person begins.

The recognition of another as a "person" is not a matter of drawing rationally watertight proofs from unassailable observational premises before that recognition can legitimately occur. Few people, if any, need rational proof to be certain that their loved ones are persons. Similarly, the distinction between "physically" created objects and personally created artifacts is often immediately apparent. The Lascaux caves mentioned in the last chapter did not give rise to the slightest doubt as to their human provenance. The cave art makes itself manifestly apparent as the work of reflective personal hands. As Stephen R. L. Clark has written,

> It is often assumed, without much argument, that all words originally have corporeal referents and are thereafter used to speak of non-corporeal experience. . . . Why . . . should we suppose our ancestors . . . had any notion of the "merely corporeal"? "Corporeality" is a piece of theory; our first words are subjective, human, and meaningful.[8]

The "impersonal" conception of exterior things and its ultimate explanatory achievements in the "natural sciences," is rather an outcome of how self-conscious and rational creatures negotiate a partial understanding of a part of the world in which they participate. Similarly, our familiarity with personal creation in the form of the artifacts from which our identity

7. Cottingham, *Spiritual Dimension*, 80.

8. Clark, *Plotinus*, 21.

as corporeal actors in the world cannot be dissociated, precedes our understanding of an "impersonal" causally and internally regulating nature.

The philosopher P. F. Strawson argued that the basic conceptual structure through which we grasp (or interact) with aspects of the world must have basic concepts around which they hinge.[9] He characterized these concepts as "sortals," those that provide the means required to "sort" the material into basic categories which any conceptual schema must have: "external events" is perhaps one of these, while "person" is another.

Without our experience of the world beginning from that division, there *is* no experience of a "world." The idea of a "person" or another rational consciousness to whom an external world appears is fundamental to *any* thought, speech, imagination, perception, and behavior; it is a basic *sortal* through which our ordinary conception of the world organizes the fundamental categories of existence; it enables the *primary* distinction in virtue of which we are enabled to distinguish between and refer to the external world, ourselves, and others.[10] It worth noting as a subsidiary point that without this rock-bottom distinction, the idea of a "science" of natural events is incomprehensible.[11] The very concept of an " idea" involves a subjective consciousness in which it appears and to which it is visible. Without a rational consciousness there cannot be a "science."

One may even suppose that our ancient ancestors got it right: there is an epistemological integrity to the initial human experience of the world and nature as indissolubly expressive of personal life in all its structures and forms, including their own corporeality. They could not, of course, have been expected to develop the marvels of scientific thought and practice within a few short years. If the history of science shows anything, it shows the enormous complexity of thought, social order, and technical accomplishment involved in the emergence of the natural sciences.

The understanding of the world as the expressive outcome of divine presence may even be a necessary intellectual preliminary to being able to understand the natural world through mathematic-empirical principles.

9. Strawson, *Individuals*.

10. Strawson's work has generated a substantial secondary literature. A recent publication providing a useful history of the concept of personhood devotes a chapter to a comparison between Strawson and Thomas Nagel, the contemporary American philosopher who argues that individual experiences cannot be redescribed in wholly objective scientific terms without leaving a first-person subjective remainder. See Rist, *What Is a Person?* Rist's book also discusses many of the major philosophical positions on personhood developed over the centuries.

11. Hacker, "Strawson's Concept of a Person," presents a fundamental critique of the descriptive metaphysics upon which Strawson bases his claim regarding the basicness of persons.

Once one understands the world as the outcome of creative agency, a path is open to conceiving it as we understand our own artifacts: intelligible structures in virtue of the purposeful and rational manipulations that bring them into being. Once we do this, we can develop the idea of "nature" and the natural sciences to go along with it.

It does not *follow* from that realization, as naturalists suppose, that we need to evacuate teleology and meaning from any structure by assigning ontological privilege to elements composing the structure. However much we come to know about the materials and composition of van Gogh's *Starry Night*, what it expresses is not thereby ontologically diminished or reduced to those composing features and substances: in fact, its analysis as a purely physical object is posterior to its identification as a painting—a teleological artifact which the various material and compositional analyses have as their focus. I return to Ridley's observation: "The artist must be seen as present in his work, much as a person must be seen as present in his behaviour, rather than as separate from it, behind it, or, above all, logically independent of it."[12]

If he is right, then a substantial consequence follows: *no reduction of a painting to its naturalist properties will ever be sufficient to express the personhood in it. Any identification of the object as an artistic display will be dependent for that categorization on the prior recognition of it as embodying the personhood of another.*

Similarly, there is nothing in the natural sciences that dismisses our experience of the personal in the causally intelligible structures in nature. To the contrary, as I have tried to show, the presence of beauty in its innermost forces is expressive of the intent and value one might expect to find in an intentionally constructed work of art.

It is in this way that we can see ourselves and the nature of which we are part as expressing the world's coming to be as an "artifact" from the hand of the divine. To perceive the world in this manner is to experience its fullness as an expression of creation. The experience shares a conceptual link with the concept of the "person" and their objects in human life: a fundamental starting point through which we organize a world.

Of course, there will be no more basic concept than that of God. In our thought about the ultimacy of the world, God would be the final resting place upon which all conceptual divisions rest and from which they flow. Questions of vast theological, philosophical, and scientific significance, I appreciate, are raised by this assertion. I cannot clearly begin to approach these within the remit of this work.[13]

12. Ridley, "Expression in Art," 218.

13. Legenhausen, "Is God a Person?," offers a dated but still relevant overview of

David Bentley Hart, discussing the Christian God, draws what is probably a critical distinction between human persons and divine personhood in the following manner. "We," he begins, referring to human persons,

> belong of necessity to an indissoluble coinherence of souls. In the end, a person cannot begin or continue to be a person at all except in and by way of all other persons.[14]

Only God, he asserts, is really "simple" and personal in the fullest sense. In his simplicity he comprises every modality of personal being. For the theist, as Bentley Hart's discussion implicates, the divine concept of the person sits at the very center of Strawson's sortal arrangements. It will be the basic end point of all and every possible thought about the origins of the cosmos and all that is extant and possible within and without it. Every conceptual division thereafter will flow from that primary source.

Like any certainty we have about the personhood of another, any certainty we have about God's person will emerge from an encounter with some manifestation of that personhood. While we might recognize the hand of divine intent in the presence of the world, to describe the divine nature is a perhaps impossible task.

Because we participate in the attributes of God's person, we may hope for some illumination from that participation by reflecting on ourselves as persons. It is, of course, one thing to refer correctly to persons and another to specify what any person might be. What determines personal identity and the metaphysics of personhood is the subject of many different answers, from the suggestion that there is no such identity but a constantly changing set of features (David Hume) to the idea that consciousness of memories is where identity resides (John Locke) or that consciousness itself "spreads out" as a unique substance across a person's life. I cannot begin to address these complex philosophical issues here. However, C. Stephen Evans offers a description of features commonly associated with the use of the term "personhood" and justifies his choice with the observation that

> in ordinary life the term is used with at least some measure of understanding, and it is possible to single out some of the concepts that are crucial to the notion of personhood.[15]

It is upon this broadly accepted account I will rely on when referring to aspects of the personal. As importantly, I will largely assume that discussion

arguments both for and against the question posed in the title.

14. Hart, *That All Shall Be Saved*, 56. See also the third meditation on p. 130 ("What Is a Person? A Reflection on the Divine Image").

15. Evans, *Preserving the Person*, 11.

of one aspect of personhood will, by and large, *entail* reference to the others. Thus, reference to "intentional" will presuppose "purpose" and "rationality," for example.

"Persons," Evans suggests,

> are first of all *agents*. They are beings who make *choices* and then act on those decisions. . . . Persons are [not] mere (purely rational) thinking machines. . . . Persons desire, love, want, wish, dislike, abhor and generally adopt a multitude of . . . attitudes towards a complex variety of things. . . . Despite all this complexity, persons are nonetheless thought of as possessing an essential *unity* and *continuity*. Not only is a person considered to be unified at any one moment; a person is usually even thought to be in some sense essentially the same person that he was in the past and will be in the future. These key concepts—actions, choices, consciousness, values, freedom, reasons, purposes, responsibility, sociality, unity—define for us a conceptual . . . word-picture . . . that we shall call the image of the personal.[16]

It is of course possible to point to a strong thread in the Judeo-Christian tradition, that any conception of the divine as personal draws on aspects of this conception, if not all of them. Meister and Drew, for example, have this to say:

> God is generally understood to be personal (or at least, not less than a person: one who possesses mind and will, has goals and plans and purposes and so on), ultimate reality (the source and ground of all things), distinct from the world yet actively involved in the world (creator and sustainer), and worthy of worship (wholly good, having inherent moral perfection and excelling in power).[17]

This definition echoes those of the human person offered by Evans. The theologian John Hick says, "Theism . . . is strictly a belief in a deity but is generally used to mean belief in a personal deity."[18] Richard Swinburne states that "God is a person . . . the creator and sustainer of the universe."[19]

The ontological naturalist of course will deny any content to the idea of a personal divinity however conceived, certainly not as ultimate, let alone

16. Evans, *Preserving the Person*, 10–11.

17. Meister and Drew, *God and Evil*, 1.

18. Hick, *Philosophy of Religion*, 5.

19. Swinburne, *Coherence of Theism*, 1.

the Trinitarian God of Christian tradition.[20] I suggest that one fundamental difference between naturalists/atheists and theists can be argued to often turn on the question of the presence of the *personal* and *teleological* in an overall conception of the nature of existence, the question of divine "design" being an instance.

Divine Teleology in the Cosmos: Naturalism vs. Theism

David Hume's contemporary and interlocutor, the Scottish philosopher Thomas Reid (1710–1796) responded to Hume's critique of the design argument in the *Dialogues Concerning Natural Religion* in a lecture shortly after the posthumous publication of that work (1779). Hume had argued that induction could have no place in justifying God's designing nature since we have no experience of God in Nature.

Hume's argument was based on his assumption that anything that was to count as knowledge of the external world could only have resulted from the habitual observation of two kinds of thing occurring regularly, one before the other. On the basis of such an "inductive observation," Hume suggested the idea that "design" could be perceived in things could not be supported.

Reid argued, in reply, that induction was irrelevant to the design argument, contending instead that we reliably infer design by identifying the "marks of intelligence and wisdom in effects."[21]

What Reid put his finger on was our strong tendency to suppose that certain features of the world around us seem to be the outcome of purposeful organization, and we perceive this "design" immediately through its effects. It is a tendency that seems to be widely shared (see below).

Can the urge to perceive God the "person" in the provenance of the necessary structures of existence and the contingent beauty of mathematical physics be better explained by a naturalist "theory," something that conforms epistemologically to the methods of science *and* the assumptions of ontological naturalism? A theory of that kind will have to account, among other things, for why it is that the language of teleology appears to have an intuitively indispensable role to play in our conception and explanation of personhood, and personal agency.

20. Van Inwagen and Zimmerman, *Persons*, consists of a series of chapters discussing different concepts of the person from the points of view of physicalism, substance dualism, and idealism. The volume is concerned to relate these to questions of divinity and debates in the philosophy of religion.

21. Reid, *Lectures on Natural Theology*, xiv.

Teleological explanations of behavior are a fundamental mode of understanding ourselves and each other. They are of such central importance for us to our lives, identity, goals, values, and wellbeing that it comes spontaneously to us to deploy teleology as a method of explanation across a variety of different types of event.

There seems to be an overwhelming tendency (see below) among *Homo sapiens* to extend teleological explanations way beyond their immediate application to other humans, to try and make sense of the world and its happenings in meaningful and purposeful terms. Attempts at comprehension often implicate "supernatural" agency to explain the behavior of animate and inanimate events of the natural world. The Greek and Roman pantheon of gods of the pre-Christian classical era is the obvious example.

The presence of supernatural powers and deities have naturally been strongly associated with health, happiness, existential anxiety, and the explanation of events that would otherwise remain random, frightening, and unpredictable. When it comes to questions surrounding the ultimacy of being and the structures of nature, we thus appear to be psychologically predisposed to teleological explanations involving the *intentions* and *purposes*[22] of divine beings.[23]

An atheist, biologist, and codiscoverer of the structure of the DNA helix, Francis Crick was very alert to this tendency. In his scientific biography, *What Mad Pursuit: A Personal View of Scientific Discovery*, he wrote, "Biologists must constantly keep in mind that what they see was not designed, but rather evolved."[24]

A clear difference between physicalist naturalism and theism thus rests on whether or not the perception of the world at *any* point in its comprehension manifests a personal divinity. The physicalist and ontological naturalist denies this is possible. Nothing personal or teleological is to be admitted into any final account of the world: reason and science are the final determinants of that possibility and a "subjective" response to the world as manifesting personhood is an illegitimate epistemological warrant for the

22. De Cruz and de Smedt, *Natural History*, ch. 4 ("Teleology, the Design Stance, and the Argument from Design"), provides extensive empirical evidence drawn from developmental psychology, cognitive psychology, and cross-cultural studies concerning the human tendency to discern teleology in nature and to make probabilistic judgments. I consider later whether this work supports a reductive analysis of theological intuitions.

23. As an example, Hutton, *Pagan Britain*, provides an extensive book-length review of archaeological evidence from the British Isles suggesting perceptions of supernatural presence. The work ranges from the Paleolithic, Mesolithic, and Neolithic periods through to the arrival of Christianity in Britain.

24. Crick, *What Mad Pursuit*, 138.

case. If this diagnosis is at least part of the explanation of the differences between them, then the motivational conflict between the atheist naturalist and theist becomes clearer.[25]

How teleology and personhood is to be understood in relation to cosmic origin will markedly differ as a consequence: the theist attributes its creation to a divine author and assumes that its structures are a consequence of his intention. In harmony with his biblical sources, revelation, and personal experience of the world as thus perceived, the theist takes it that God is not only the creator of all that is, possessed of the qualities of omnipotence, omniscience, and perfection, but also a "loving God." His intelligence and personal characteristics are inextricably "one" in his nature.[26]

The atheist or naturalist, dubious about the explanatory value of such attributions and bearing in mind their demonstrated empirical failures in the natural world (so she takes for granted), will seek an explanation for the human tendency to see intention and the personal in nature and its origins elsewhere.

Firm in his conviction that the successes of the sciences, formally bracketed by methodological naturalism, provide support for his metaphysical naturalism, the atheist insists that the cosmos is blank. No evidence of the personal or intentional activity is to be found or perceived within it. No meaning, symbol, sign, argument, or anything expressive of mind or life has been excavated by the sciences in its explanatory innards, he insists. One consequence of this is rather nicely expressed by Mark Vernon. Discussing the scientific enlightenment of circa five hundred years ago, he writes,

> Immanuel Kant declared "dare to know" as the clarion call for the scientific age. However, this daring has led to many troubles taking hold of the Western soul. A key one is the so called "death of god," the widespread sense that, in truth, we may be drifting through empty meaningless space, both literally across the cosmos, and metaphorically in our minds.[27]

25. The reference to "conflict" here is not intended to suggest that the relation between science and religion must be understood in such terms. The conflict, as I understand it, is specifically between ontological naturalism and theism, as interpreted by the former. See Plantinga, *Where the Conflict Really Lies*.

26. I express it this way for the sake of brevity and acknowledge the theological and philosophical complexities and difficulties associated with it.

27. Vernon, *Secret History of Christianity*, 5.

Naturalism: Explaining the Illusion of Divine Teleology?

The atheist naturalist must accommodate the strong impulse we have to see design or supernatural forces at work all about us. Since it cannot be for him a perception of *actual* design, it must be a consequence of our internal cognitive workings. A projection, perhaps, of our strong built-in cognitive inclination to interpret nonanimate behavior via the categories of the personal, or the consequence, perhaps, of deep psychological yearnings projected onto an empty physical canvas.

An atheist naturalist, like the philosopher and cognitive neuroscientist Daniel Dennett, might even acknowledge that the inclination to discern the hand of design emerges from its predictive usefulness in negotiating not only ourselves but the world around us. Dennett calls this taking up the "Intentional Stance."[28]

Combined with some evolutionary story about the usefulness of the Intentional Stance in self-preservation, Dennett's argument, or something like it, is useful to the naturalist in several possible ways: Firstly, it seems to explain why we feel so strongly drawn to intentional explanations of phenomena even when we have confirmed nonintentional explanations of how they work and came to *look* designed; our cognitive and affective structures embody a tendency that is a component of our "deep" evolutionary heritage. Secondly, it can be mobilized to provide an alternative and well-warranted explanation for why we are so strongly attracted to supernatural explanations.

Explanations of this sort generally run something like the following: our inherited tendency to attribute rational purpose to nature around us leads us to ascribe rational personhood as an ultimate explanation for the presence of the natural world around us, *even* when we are familiar with the scientific explanations. In this sense our "intentional tendencies" are irrational.

This reductive account explains our "irrational attachment" to supernatural belief. The naturalist might be persuaded that our perceptual and experiential intuitions of supernatural teleology are a neurological outcome of evolutionary events; the subjective authority of that emergent experience might even endow it with a compelling force urged upon us by evolutionary cognitive and affective neurological structures beyond our rational control. That instinctive force might engage one even while it is recognized that it lacks any epistemological warrant.

28. Dennett, *Intentional Stance*.

"Science and reason," of course, provide the bulwark against the apparent epistemological power of our intuitions. We "perceive" divine personhood when it is not, in fact, there. Consequently, intelligent forethought, love, or beauty are not present in any aspect of the universe's construction, in the reality of its appearance to us, and certainly there could not be any purpose in its progress through space-time.

The theistic attachment to a God-given eschaton is an illusion. If we make the attempt to find traction for the concepts of the personal anywhere in the universe, we will be guilty of a primitive usurpation of the truth-detecting functions of the sciences by an equally primitive supernatural animism. For the ontological naturalist and the atheist, this represents a reprehensible "prescientific" apprehension of the universe, a nonexplanatory and unnecessary ontological extravagance.

In the next section I inquire how the scientific disciplines can help with deciding between the options of a personal designing god responsible for nature and the skeleton case I have laid out above in support of the reductive naturalist claim proposal. The essence of the question I ask is, "Can it help at all?"

The Explanatory Limits of the Human Sciences in Questions of Divine Attribution

Why *do* we spontaneously and intuitively suppose that complex items display order and design that requires a designer or designers? The discipline of the cognitive science of religion, drawing on the experimental techniques of psychology and the social sciences, has attempted to answer this question. Among its experimental findings much evidence appears suggesting that aspects of belief in the supernatural are common to all human populations. The most common of explanations for the suggestion that features of belief are invariant among *Homo sapiens* are those that take their stance from evolutionary theories of origin and development.

This latter stance has given rise to much debate. Paul Bloom offers a succinct summary of the belief that "religious" belief is an outcome of the evolutionary development of human cognition. In his essay "Religious Belief as an Evolutionary Accident," he hypothesizes that

> religion is a non-adaptive by-product of "universal cognitive biases," including hypersensitivity to agency, a natural propensity

> to see non-random design as caused by an intelligent designer, and body-soul dualism.[29]

Daniel Dennett's picture of the "intentional stance" is an instance of the view that we have "universal cognitive biases" to teleological explanations.

De Cruz and de Smedt divide the capacities of cognitive systems that are presumed to emerge from evolutionary pressure into two categories: *adaptions* "which deal with specific adaptive problems and *byproducts* . . . which do not fulfil adaptive functions but emerge as byproducts from adaptions."[30] Bloom, for example, suggests that religious beliefs, while not being directly adaptive, nevertheless result from the normal operations of the cognitive system.[31] In this sense they are "natural."

De Cruz suggests that certain possible explanations follow from this: "Religion may be a product of our tendency to discern agents in the environment, it may be a product of our tendency to see teleology and design in the natural world, or it may emerge from the intuitive distinction we draw between minds and physical objects."[32]

There is, of course, a third option: that is that "religion is natural" because there *is* a God to be encountered in nature. Can the cognitive literature help us to finally decide which of all these hypotheses are the likely

29. Bloom, "Religious Belief," 124. Bloom's chapter occurs in the collection of essays in the book immediately above both for and against the hypothesis that belief in a deity or deities is a spandrel of human evolution. Several ways this might be the case are considered, from D. Johnson and J. Berings's contention that human systems of intentionality and the threat of supernatural punishment interact to force mutual benefits in social systems ("Hand of God, Mind of Man"), to J. Barrett's discussion of the ontogeny of cognitive mechanism predisposing to religious belief.

30. De Cruz and de Smedt, *Natural History*, 22.

31. Bloom, "Religion Is Natural."

32. De Cruz and de Smedt, *Natural History*, 22. Examples of the explanatory power of intentional rationality in accounting for nature and answering questions of ultimate existence may be found in the ancient Western world and in cultures predating what we now call the natural sciences. The design argument, drawing an analogy between artifacts and natural order, extends in the Western tradition back to Plato's *Timaeus*, Socrates, and Cicero. A version of the argument appears in Cicero's *De Natura Deorum*. Related forms of reasoning also appear outside the Western canon. The Hindu philosopher Śaṅkara, for example, embraced the idea of a designed world. De Cruz and de Smedt argue that the design argument reached its greatest popularity between the seventeenth and nineteenth centuries, with Paley's *Natural Theology* among its best-known expressions.

Not all religious belief, however, entails commitment to a transcendent deity. Livia Kohn, a scholar of Chinese religion, for example, contrasts Western monotheism with Chinese religious conceptions in which "the ultimate in Chinese religion is a process of realisation and experience, part of the world yet not accessible to worldly means and thus the opposite of the Western concept of God." Kohn, "Ultimate Reality," 32.

ones? There is a reason for skepticism about the possibility that the findings of the cognitive science of religion can offer a final verdict.

What I mean is this: If all the literature from the cognitive psychological investigation of religious belief showed without question that *Homo sapiens* tended to perceive the supernatural in some form, what would this mean?[33]

Both theist and naturalist would claim it as proof of their respective positions: the theist because scientific support for universal intuition would support an epistemologically secured response to the theistic design of the world. In other words, we tend to perceive God (or gods) in the world because God is present there.

Variance in understanding that presence could be easily explained by the proposal that like any intellectual endeavor, understanding the divine nature of what is before us is a long, hard process. Even our knowledge of what constitutes a human person and their workings, including oneself, remains a task which can often appear dauntingly unattainable, so understanding divine presence in relation to the world could hardly be supposed to be any easier.

The ontological naturalist or atheist, on the other hand, would interpret any evidence from cognitive mechanisms, social psychology, and a plausible evolutionary theory as indicating the human origins of the idea of a god. As in the case of Dennett, the ontological naturalist would no doubt maintain that the persistent perception of intentional design is an outcome of other adaptive capacities. From the ontological naturalist's perspective, the step forward in knowledge has already been taken anyway, leaving behind the theist thesis as a hypothesis refuted.[34]

33. Ruse, *Evolution and Religion*, provides an interesting and balanced presentation of arguments on both sides of the debate in the dialogical form suggested by its title.

34. I should note that I am not suggesting that disciplines investigating the cognition of religious belief have nothing to contribute to our understanding of how human beings are cognitively and neurobiologically structured so as to be inclined toward perceiving divinity in the world. Such investigations, however, would no more demonstrate that God does not exist than the existence of dyscalculia—a neuropsychological disorder that makes even basic arithmetic difficult—would demonstrate that mathematics is merely a product of the brain regions involved in numerical cognition, or any other psychological dysfunction consequent upon brain disorder.

PART 2

The Divine Teleology of Metaphysical Structure

The classical nineteenth-century argument for design,[35] that of William Paley[36] is perhaps classically and typically associated with the final failure of intimations of perceived design in nature. Paley provided a now iconic example to illustrate his argument: the picture of someone coming across a watch among the flora of a heath.

The most natural of inclinations, he proposes, is that it should be taken to be the product of design. The intuitive pull of this explanation of the complexity and functioning of the biological organisms of the natural world was, of course, apparently refuted by Darwin's theory of natural evolution[37] which (apparently) dispensed with the need for purposes and final causes in biology.[38]

The displacement of teleology and "design" does not, however, subtract from the attraction of the idea that design has an appropriate sense-making contribution to our overall comprehension of reality. I have already argued in previous chapters that, in any event, teleology has a role to play in understanding the artifact and behavior of natural creatures including ourselves. I also argued that teleology lies at the heart of the possibilities of the natural world.

35. The "father of chemistry," as noted earlier, defended, or at least left open, the possibility that final causes might continue to play some role within the experimental sciences. I mention this again to indicate that the supposed "failure" of the design argument was not a straightforward deduction from the early successes of empirical science. See Boyle, *Disquisition About the Final Causes.*

36. Paley, *Natural Theology.*

37. As laid out in *On the Origin of Species.*

38. I have already discussed Deacon's argument for the incorporation of teleology into biology. Another biologist who has defended a related position is Francisco J. Ayala, professor in the Department of Ecology and Evolutionary Biology at the University of California. In an essay titled "Darwin and Teleology," Ayala argues that teleological explanations remain necessary within biology while nevertheless being fully compatible with physical explanations of the structure and behavior of organisms. His position does not involve recourse to intelligent design theories, which maintain that certain forms of biological complexity cannot be adequately explained within an evolutionary framework. See Ayala, "Darwin and the Teleology."

Feser, in a comprehensive work titled *Aristotle's Revenge*, argues that many of the fundamental categories of Aristotelian philosophy of nature—including final causation, actuality, and potentiality—remain implicitly presupposed by modern science.

Since teleology is closely conceptually tied to design, one nonreductive possibility is that the intuition that design is found around us is well founded.[39]

The perception of design, it might be argued, is not illegitimate but *wrongly sought* in the domains properly belonging to the sciences. It should be noted here that divine "design," as I use it here, is but a shorthand for the idea that *however* nature functions, it is the outcome of divine activity. Of course, this raises questions about God's intentions for the universe and how they are to be realized (if for example, many processes are random or outcomes only possible rather than determined), not for the question that their operation was "designed" to be such in the first place.

To recognize that design has a referential and meaningful arena of application is not necessarily, as might be thought, an attempt to save a concept displaced by empirical advances in understanding the natural world. It might represent an appropriate intellectual and theological shift in our overall comprehension of reality to which the natural sciences have contributed.

The contribution of natural science to "design" is to help shift us to the understanding that the *perception of design is a response to the metaphysical structure of things that we develop in coming to understand the world through our empirical and philosophical engagement with it.*[40]

I have already offered reasons, for example, to suppose it might be plausible that that structure involves elements congenial to the very idea of a lawfully accessible nature as involving design.

Any "law of nature" must describe and predict what happens. This is much the same as saying that any law of nature must have the potential to describe the changes that regularly occur in the phenomena in which it has an interest. It is to say of any law of nature used in explanation that it *must* yield outcomes appropriate to these tasks if it is to serve theoretically useful functions.

39. Augros, *Who Designed the Designer?*, offers a highly accessible Thomist response to the objection that, if the universe requires a designer, then the designer must himself require a designer, thereby initiating an infinite regress. Augros argues that the regress cannot simply terminate with the universe itself. Although the book is not written as a strictly academic work, its presentation of both the objection and the Thomist responses is notably clear.

40. I put it this way because the traditional idea that we "abstract" metaphysical principles from our sensory and cognitive engagements with the world does not quite capture the picture I want to elucidate: that our very experience of that world as yielding physical objects and sensory creatures to perceive them is a function of the metaphysical principle that make possible that experience as one containing an "external world" and "experience."

A law of nature presumably has the descriptive and predictive capacities it has because the principle on which it is based derives from metaphysical assumptions, that, for example, nature is "uniform"[41] or "causal processes are regular," or "all objects come in structural forms with general properties that flow from that necessary principle." As I discussed, for example, in the case of mathematical physics, the equations that describe the fundamental forces (e.g., electromagnetism) are themselves constrained by the aesthetic principle of symmetry.

The personal agency *behind* the metaphysical principles and conditions that determine the contingent nature *of* any physical structure and the consequent concept of laws of nature is where design may be located free from the threat of attempting to displace empirical theory. Inasmuch as evolutionary theory is about one aspect of one contingent structural process of the natural world, it shares the same metaphysical determinants and presuppositions. This thought is not one that emerges in ignorance of a scientific solution but comes out of reflection on the conceptual prerequisites of such solutions.

The thought is that "design" might lie behind the metaphysical tapestry which provides for the discernment of the idea of an "empirical nature." As I argued in the first chapter, and reiterated at the beginning of this one, the concept of an empirically approachable independent "nature" is not a given of a perceptual encounter with things around us but an intellectual achievement of human persons who are part of the world. In that achievement we uncover something of the "design" of natural reality.

The aesthetic manifestations of natural structures, for example, are not comprehensible outside the broader picture, *and* serve expressive purposes. The perception of beauty exemplifies this: rich in provoking desire and hope, it is intrinsically evocative of "meaning," whether in physics, in other aspects of "nature," or in human artifacts.

It is plausible to offer the thought that behind the provenance of aesthetic structures at the heart of matter lies the divine personal. The understanding of metaphysical structures, inclusive of those providing for the laws of nature and the teleological, seems to be one that can engage a cause beyond both. In the Judeo-Christian tradition this is of course a personal God.

No less a figure than Charles Darwin acknowledged this. After reflections on the infamous case of the Ichneumonidae, a wasp which places its

41. By "uniform" I refer to a metaphysical precondition, not physical occurrences: if our understanding of biology yields local differentiation even within local populations, then the fact of our being able to articulate it as an occurrence which occurs and reoccurs is a mark of uniformity.

eggs in the living body of a caterpillar to provide its offspring with living flesh to feast upon, he wrote to his friend Asa Gray, a prominent biologist of his day, with his ethical reflections on evolution.

Darwin, quite reasonably, argues in the letter that the apparent intrinsic cruelty manifested in the Ichneumonidae's behavior could not be the plan of a "beneficent creator." Despite this thought, he conceded to another of equal conviction:

> I cannot anyhow be contended to view this wonderful universe & especially the nature of man, & to conclude that everything is the nature of brute force. I am inclined to look at everything as resulting from designed laws, with the details, whether good or bad, left to the working out of what we may call chance.[42]

"Beauty"—like good comes with bad—comes with ugliness. This apparently necessary combination of evaluative opposites is a perennial source of theological concern. It should not be a surprise that Darwin struggled with the two thoughts: the necessity to attribute design behind the very presence of evolution and the disregard of evolutionary processes to suffering.

Despite his struggle to reconcile what seems to be a blatant cruelty with a good God, this was clearly not enough to inhibit the intrusion of the concept of design into Darwin's thinking about the place of evolution in the broad scheme of things. Curiously, this comment is often overlooked but is rich in exegetical possibility. What he presents is, in effect, a *metaphysical* reflection on how the mechanisms of evolution that work via what he says is "chance" ever came to be.

It is of similar interest that he precedes the use of this word with the phrase "what we call." The first of these reflections suggests that he was aware that evolution should be placed in a broader context in which the *fact* that the slow coming into being of forms of life works the way it does, via natural selection, survival and "chance," is something that itself requires an explanation. What or who enabled the laws and mechanisms that underpin the principles of evolution? The answer he is compelled towards is that behind the coming into being of natural "law" and evolution of creatures is design.

This interpretation receives further support from the second point. The use of the phrase "what we call" leaves the door open to precisely *this* metaphysical supposition. In other words, while correctly showing that evolutionary mechanisms do not predict the kinds of creatures that will evolve

42. Darwin to Asa Gray, May 22, 1860, quoted in Ruse, "Argument from Design," 23. For a fuller discussion of the correspondence between Darwin and Gray, see Lennox, "Darwin/Gray Correspondence." See also Gray, *Letters of Asa Gray*.

by chance, *that* creatures *can* potentially evolve might be designed *into* the universe *ab initio*.

This point is emphatically *not* the claim that if we pursue the explication of natural structures and events haltered by physicalist strictures, we will inevitably encounter designed elements *in* the universe whose *complexity* cannot yield to the materialist causes assumed by the natural sciences. This is the wager of intelligent design theory: that complexity is to be found in certain natural objects, perhaps cells for example, or other biological functions and structures that cannot be explained by the vagaries of natural selection. Design is, however, *already* present in the universe *before* the identification of entities that are claimed not to have originated by natural processes. It is a condition of there being anything at all.

Creation, Design, and the Natural Sciences

The more careful claim draws on the distinction between Creation and Universe. The whole panoply of laws of nature, principles of development, mathematics, the potential for sentient creatures to evolve, space-time, regularity, and so on are, one might say, "created" or brought into their various forms of "being." The functioning of that universe and "how it goes," as Cardinal Baronious (1583–1687) is reputed to have said of the heavens in the Galileo case, is a matter for the sciences. This does not, of course, mean that there are no conceptual connections between the coming into existence of the universe and its functioning.

Physicist and theologian Stephen M. Barr articulates this in an essay titled "The End of Intelligent Design." Discussing his opposition to intelligent design theory, he writes,

> The older (and wiser) form of the design argument for the existence of God—one found implicitly in Scripture and many early Christian writings—did not point to the naturally inexplicable or to effects outside nature, but to nature itself and its ordinary operations—operations whose "power and working" were seen as reflecting the power and wisdom of God.[43]

In another article by Barr titled "Chance, by Design," he fleshes this out. His argument there is directed at the question of "chance" as understood technically within probability statistics and its use within natural sciences concerned with biology and evolution.[44] Barr writes,

43. Barr, "End of Intelligent Design."

44. Sober, *Design Argument*, provides a technical discussion of intelligent design

> When biologists start making statements about processes being unsupervised, undirected, unguided, and unplanned, they are not speaking scientifically. No measurement, observation, or mathematical analysis can test whether God planned a development like a genetic mutation. What apparatus would one employ? "Being unplanned by God" is simply not a concept the fits within empirical science. Being "statistically random" on the other hand, is, because it can be tested for.
>
> And suppose we did define Neo-Darwinism to include the belief that the world is "unguided and unplanned" by God. Then Neo-Darwinism would be atheistic simply by definition. Only if Neo-Darwinism is defined in terms of its strictly scientific content does how well it comports with a religious view of the world become and interesting and debatable question. This is the question I shall now examine, particularly how the kind of randomness posited by Neo-Darwinism theories can be consistent with divine providence. But first we must understand what providence and randomness are.
>
> Most traditional forms of biblical religion agree as to what providence is. They may dispute how providence relates to human freedom, but their differences don't affect their understanding of merely physical processes such as evolution. The traditional doctrine is that God, the transcendent source of being, wills by one timeless act that all things exist, wherever in space and time they do exist. He is the creator of all finite beings in every aspect of their being, and hence he creates them with all their nature potentialities, powers, and relationships, including their causal relationships to each other.[45]

In the last paragraph, Barr locates the proper domain where questions about design properly belong. Owen Gingerich, the astronomer and historian of science at the Harvard Smithsonian Center for Astrophysics, in the book based on his 2006 Noble lectures, *God's Universe*, has a chapter with the pointed question, "Dare a Scientist Believe in Design?" that develops the same point. "Is the universe designed?" he writes,

> is not a scientific question. . . . The reason is simple. The answer is one without an answer in the scientific sense. It is a

and the fine-tuning argument. He also incorporates a useful discussion of probability theory relevant to analyzing such arguments when formulated in probabilistic terms.

45. Barr, "Chance, by Design."

> metaphysical one whose answer will come out of metaphysical reasoning.[46]

Later, when answering the question underpinning his chapter, he acknowledges that to believe in a designed universe requires accepting teleology and purpose, "and if that purpose includes contemplative intelligent life that can admire the universe and search out its secrets, then the cosmos must have properties congenial to life. For me, part of the coherency of the universe is that it is purposeful."[47]

The invocation of design clearly implies the presence of a "designer," and the implication of that is the presence of the personal. Behind the distinction Barr makes, and Gingerich's quoted comments, are all words—"planned," "creator," "unsupervised," "wills," "purpose"—that encompass the mindedness of personal agency.

Whatever we are to make of the personhood of God, and whether or not we can successfully refer to him and the extent to which our conversations about his attributes are captured by the common languages of our own human personhood, are questions that I cannot fully address here. What I can insist on, nevertheless, is that any theist account of God's nature or argument for God's existence and his bringing about of our world will embrace *somewhere* in its construction reference to the personal as we roughly understand it and make use of it in the ordinariness of human life.

Since this being cannot itself be caused *by* the physical world it creates, and since that world cannot be the cause *of* itself, the term "creation" can only gain any sense here from its semantic relation to the idea of a being who *intended* it to come about.[48] The genesis options are binary: either personal or impersonal. I discuss these observations in greater depth in the next chapter.

It is sometimes suggested that this limits what any such argument from the existence of the cosmos can show is the outcome of the actions of an abstract "superintelligence." "Superintelligence" in this context suggests

46. Gingerich, "Dare a Scientist Believe in Design?," 70.

47. Gingerich, "Dare a Scientist Believe in Design?," 77.

48. What has been called "Hume's Objection" insists that if we have a series of causes, such that A causes B, B causes C, etc., then, if we can explain the causal relations between each two elements across the series, there are no further explanations to offer—i.e., of the Series considered, if it can be, as a Whole. This argument relies on the analogy between "the Series" as an object and "the Universe" as an object. It is not clear, however, that the coming into being of the Series consisting of the chain of causes does not require some account of how *it* came into being, and that that Series does not require a final cause to support it.

an abstract rationality lacking the kind of motivations that move persons, those of value and feelings.

"Reflective rational intention," however, is a teleological attribute of persons. And what any intention is must be spelled out by reference to the reasons for having it, the desiring of its outcome, perhaps because it is good, and acting to bring it about. Hence, my minimal claim that whatever brought the universe into existence must have the property of sharing the agency of the personal. However the reference of that term is to be understood when applied to the divine,[49] it must on pain of meaninglessness have this minimal content.

In the following sections I offer examples to illustrate and support the point. The first of these is drawn from questions about the origins of the universe; since these reflections involve mathematical physics and early cosmology, they are strongly related to questions of ultimacy, creation, aesthetics, and personhood. The second example comes from the ontology of morality. I choose this because of its particular normative resonance for personal choice, agency, and free will.

Cosmological and Design Arguments and Evidence of the Personal

Modern cosmology appears to point to a universe "fine-tuned" for life[50] within a remarkably short time of its coming into existence. This is the "anthropic fine-tuning argument,"[51] the view that several conditions had to be present together at the "Big Bang," the birth of the universe, for life to emerge. The probability of their co-occurrence by "chance," the argument runs, is so vanishingly small that the best explanation is that they evidence the prior intentions of divine intelligence.

49. A further argument that cannot be made here might suggest that "intentions" are ineluctably bound with what is a "good" or "bad" intention, hence introducing at least the possibility that intentions are open to moral assessment.

50. Design arguments take many forms, ranging from probabilistic arguments comparing the likelihood that certain features arose by chance with the likelihood that they were designed, to arguments of a more a priori character. A useful collection of technical essays on these issues is contained in Manson, *God and Design*. See also the sections on the cosmological and teleological arguments in Moreland et al., *Debating Christian Theism*, each of which presents arguments both for and against the classical positions.

51. Collins, "Argument from Physical Constants," 89. See particularly Lewis and Barnes, *Fortunate Universe*, which provides a comprehensive philosophical and scientific assessment of the evidence both for and against fine-tuning.

It is the observation of the co-occurrence of these features at an inconceivably early stage of the universe, as well as the lauded extraordinarily low probability of various physical constants, that is important in suggesting design. Put at its simplest, if the universe were designed, it is at the point of creation where one might expect to find precisely those features that would yield the potentiality for the full flowering of the universe and *all* its inhabitants at some point in its future.

The early laying of the foundations for a house is a function of the architect's overall conception of the building, and it clearly restrains and provides foundations suited to what will later emerge. The dwelling has, in other words, been designed. The difference here, of course, being, as Aquinas insisted, that God gives being *and* essence to those elemental components that bear cosmic and life-developing potential.

That the universe *must* have consisted of structures and forces which had the *potential* for flowering into nature and life, including ourselves with our capacities for reason and thought, follows from the fact that the universe *now* contains nature and life. The application of the logical rules of modus ponens (method of affirming)—If A, then B; A, so B—and modus tollens (method of denying)—If A then B; –B, so –A—can describe this argument formally.

Let Tn+1 = some given time after the birth of the universe, Tn.

Let A = There exist properties conferring potential for life and nature at Tn+1.

Let B = Any and all properties of life and nature which exist at Tn+1 have properties which are an outcome of emergence (water consists of H_2O, hydrogen is formed earlier in the stars, and so on).

It follows logically that if A is true and if B is true, then if we deny B, we must deny A.[52] It is conceivable that someone may wish to deny B, but then they deny a cardinal tenet on which every physical science relies. Put at its broadest, this is the principle that everything in nature has a preceding

52. So, for the existence of any property *x* which does not exist at the time of creation, the universe at creation must have had the *potential* for that property *to* emerge *prior* to its emergence (If A, then B). If we deny that the universe at creation had the potential for that property to emerge (–B), then, by modus tollens, we must deny the presence of that property now (–A). But that property does exist (A), so the universe must have had the potential for its emergence at creation (B). Our only other option is to suppose that new properties or substances, or even extant properties or substances, can emerge that are *entirely* unrelated to what already exists. This cannot be a happy conclusion for the physicalist or the naturalist. It supposes the possibility that an entirely novel physical object might suddenly appear before me without any derivable physical causes.

set of natural events that explains its physical origins and its current causal functioning.

Note, en passant, that with a denial in place, the very assumption of evolutionary theory, that animate life develops from the relation between the physical environment and the physical organism, would be thrown into potential doubt.

The argument, of course, does not and *cannot* establish the physical nature and mechanisms by which the physical universe emerges, develops, and functions. Like the architect and the building, however, it can provide support for an argument that for there to *be* any such thing as a physical thing, event, or structure, a metaphysical designer provides the terminus[53] for an explanation that cannot be provided from *within* the physical world.

Design and Scientific Progress

To elucidate the idea that the understanding of the right place to assign design in our ultimate comprehension of the world is part of a greater intellectual progression to which our advances in science contribute rather than detract, as much current orthodoxy (at least naturalist orthodoxy) insists, I want to introduce the idea of "epistemological progressivism." By this I mean the idea that as knowledge of the cosmos advances, so any questions which seem at present to fall without its scope are often incrementally degraded. Evolutionary theory is an example here, and so is physics.

Theists might often consider themselves on the epistemological back foot in this regard. The cautionary tale for the theist is to be careful of a "God of the gaps" argument: using God to fill some explanatory gap in the sciences. There is another way of conceiving scientific advances, however, that can contribute to a greater comprehension of the divine personal in relation to the natural sciences.

Some scientific advances seem to enhance our aptitude to perceive more precisely where evidence of intention is to be found *without* posing it as an alternative "scientific" explanation. Consider, for example, how someone unacquainted with the very idea of a combustion engine, from a prescientific culture, might approach an explanation of a moving vehicle seen for the first time. They are likely to draw on explanations that involve perhaps "hidden horses," or concealed people who intentionally provide the moving power.

Suppose the explanatory climate of that culture gradually shifts as it comes to the idea of contemporary science (as indeed Western culture did).

53. "Terminus" refers here to logical not physical or temporal terminus.

Coming to understand the various engineering and scientific principles behind the engine, for example, does not obviate the entire framework of application of teleological explanations, but refines the domain of application of that understanding.

Increasingly, its use becomes more refined and precise, being confined to explanations of agency involving human movements of objects for some purpose and, more generally, refined to explanations of personal life and behavior. How we come to understand the place of divine teleology in relation to the existence of the "physical world" may involve similar progressions in intellectual comprehension.

Here, as a contemporary example, is an opinion expressed by Robin Collins, an Oxford philosopher, on the theistic explanation of the birth of the universe, its creation.[54] He points to several features of the very early explosion of matter/energy, and space-time coming into existence at the "Big Bang." Collins points to the degree of fine-tuning necessary for the "cosmological constant" and the initial distribution of mass-energy in the universe. In the former case the figure is one part in 10^{120}—that is, 1 followed by 120 zeroes. In the case of the latter example, the figure is even larger.

The inexpressible size of these improbabilities gives rise to the suggestion that they must have arisen by intentional manipulation, the "fine-tuning" to which I previously alluded.[55] The physicist John Leslie attempted to enunciate why the personal explanation seems such a powerful alternative. He asks us to consider a firing squad composed of many men that misses its unfortunate target, despite it being observably a near impossibility. The most reasonable assumption, he suggests, is that the members of the squad *intended* to miss their target. All other explanations fail to rise to the plausibility of this one.

Considering the enormous improbability of the figures he quotes occurring by chance, Collins says, "Cases of fine-tuning should be sufficient to show that . . . the issue of fine-tuning is not likely to be resolved by physics."[56]

He says this because he finds the inference to a designing mind *behind* these physical and structural improbabilities an almost irresistible resting

54. Collins, "Fine-Tuning Evidence Is Convincing." For a contrasting response, see Stenger, "Universe Shows No Evidence," in the same volume.

55. One popular alternative is the "multiverse hypothesis." Expressed very briefly, it argues that since there are any number of universes that come (or can come) into existence, the probability of one being "fine-tuned," as ours, is at least probable, thereby diminishing the force of the design argument. As mentioned in my chapter on mathematics and physics, one objection is that the hypothesis that there are other universes cannot be tested. I noted in this chapter that this objection has been raised by S. Hossenfelder, herself a physicist.

56. Collins, "Fine-Tuning Evidence Is Convincing," 44.

point for making their appearance intelligible. One can understand why Collins says this.

One interpretation of Collins's conviction is to argue that there being such things as scientific facts *at all* requires understanding. That understanding is comprehensively and intelligibly explained by a different *kind* of explanation, appealing to the *intention* that the world should be such as to *contain* scientific facts. Scientific facts must bottom out somewhere.[57]

An explanation for the existence of scientific facts, if there is to be such a thing, and whatsoever they may turn out to be, cannot itself be *composed* of scientific facts. The "fine-tuning" argument may be the place that scientific facts in respect of the origins of cosmology come to a particular kind of end.

This is of a piece with the frequently repeated contention of this work that the natural world as we have it requires a metaphysical context to give it its physical nature. The very concept of a "scientific fact" is not itself the production of any scientific activity. It is rather a piece of the metaphysical glue that holds together what is meant by the *achievement* of a supported knowledge claim resulting from experimental and theoretical activity.

So, one may think of expressing Collins's point this way: That there *should* be scientific facts quite as extraordinary as those of "fine-tuning" suggests a metaphysical architect. That they should be *beautiful* facts puts the final theistic icing on the metaphysical cake. Collins makes his own case this way:

> Even if physicists found a theory that entailed the initial conditions of the universe and the constants of physics fall into . . . the permitting range, that would still involve extreme fine-tuning at the level of the laws themselves.[58]

This point has an echo of that made by Gingerich above. It is different from a scientific versus a theistic explanation for natural phenomena. It could turn out that certain empirical theories of physical origins find an end at this point. Even the naturalist might be willing to concede the concept of an "end point" while denying on philosophical principle an end to the possibility of empirical endeavor. After all, for all naturalists the material structure of things is the end point of all that there is or could be. If this should be so, the fine-tuning argument would be considerably enhanced.

57. Notice that the argument that they may *not* bottom out does not invalidate the asking of the question "Why?" To which one answer may be, "God *intended* it to be so." The intentional explanation would not disappear but find its correct place elsewhere.

58. Collins, "Fine-Tuning Evidence Is Convincing," 44.

Whether or not Collins has placed the metaphysical explanation at the *right* place is a matter for ongoing interdisciplinary debate. Should there be a suitable explanation from physics, then Collins is wrong, but if this is the case, then those mechanisms would be the first point of designed origin.

The argument is a metaphysical not a physical one, so this is not an argument that can dictate conclusion to the cosmological sciences. Whether it has any defendable traction, my point in exemplifying it is to note that the pull of a teleological explanation located at the metaphysical level is hardly undercut when faced with sophisticated physics and cosmology, but may be *exacerbated* by it.

We may be cognitively simply unequipped to ever approach the perfections of divine design in its manifestations in cosmology or other sciences, but "signs" that can be understood as manifesting teleology in nature *may* be available to us in the sciences as elsewhere. But perhaps I may conclude with a final reiteration: the very idea of the sciences *assume* teleology; that science has a purpose which can be fulfilled, that of acquiring empirical explanations.

That such a possibility exists is no surprise to the theist since purpose lies at the heart of the concept of *metaphysical* creation. The world was constructed ab initio with the purpose of providing for the functioning of natural structures and the potential for the emergence of creatures who could reason their way to understanding its structure and the provision of a world uniquely fitted to be thus understood.

Ethics and the Personal

Another point to be drawn from the Darwinian morsel originally quoted is his use of the phrase "wonder" and the intended or accidental admission that "good" and "bad" might be found in the very nature of the physical world. It is the apparent presence of such events as the behavior of the parasitic ichneumonid that seems to force this judgment from him.

Darwin cannot, his naturalistic explanation notwithstanding, inhibit the desire to ethically reflect on the consequences of evolved behavior. Not only that, but he is clearly repulsed by the activities of the ichneumonid and cannot inhibit his moral repulsion to it. The question is why this should be so. There are many potential explanations of this commonly shared reaction to the cruelty displayed by most creatures, including ourselves.

Natural evil, that which emanates from purely physical objects, draws similar ejaculations of judgment even with the knowledge that intention cannot be involved. Consider responses to natural disasters, for example.

There is often a sense of the unfairness of random and inexplicable death and injury, coupled with an inchoate revulsion at the physical events themselves, even while acknowledging a simple "occurrence of nature."

Now while the question of why a good god would allow for or create evil is one of supreme importance for theistic traditions, there is a question that precedes it. The question is from whence our knowledge of and capacity to identify and condemn evil derives.

These are questions that involve theodicy and moral philosophy. They have ranged from the denial of the coherency of any objective moral realm that determines value in favor of subjectivism, to the positing of an abstract Platonic realm of ethical forms, to the suggestion that moral propositions, while sharing the form of a factual claim such as "eating people is wrong," are nothing but a disguised attitude of distaste with no content, sometimes abbreviated as "boo to eating people"—the use of the word "boo" indicating that no moral proposition has any rational or assessable content despite appearances. None of these questions can be properly engaged with in this work, beyond noting that for the theist, anyway, an inarticulable disapproval, when pertinently delivered, may nevertheless carve God's moral intentions at their correct joints.

Rather, I should like to draw attention to why it is that the moral argument for God as the *giver* of moral law to be *found* in the Universe has had a powerful appeal for many. I suggest it shares with the other classical arguments a similarly powerful attraction to elements of the personal. This is not to defend the position but to indicate the *source* of the power of its psychological appeal.

Moral assertions are, taken a face value, concerned with the rightness or wrongness of actions or thoughts that concern the sentient and the animate. A proposal that it would be wrong to chop up a wooden table that located its force as something to do with the *physical properties* of the table would not be taken to be a matter for debate; it would simply be recognized that a wooden table is not an appropriate entity at which moral concern should be directed because no sentient properties are attributable to it.[59]

59. The argument here is that "being wooden" is not itself sufficient for being an object of moral concern. What would have to turn something wooden into an object of moral attribution is a property that derives from outside its "woodenness." Thus, it may be an object of cultural or personal significance. Environmentally, of course, the threat that wood might disappear is an object of concern, but this is not because it is wood per se but because of value attached to it as an item of environmental value.

It may be argued that environmental concerns obviate this argument. However, environmental arguments do not suppose that cruelty or suffering is being imposed on insentient matter but rather that it is a good-in-*itself* that the environment be preserved. Quite why this should be so, if the world is blind and dark to itself and has no intrinsic

Human persons, sentient creatures, and—more contentiously, at the boundaries of what is to count as sentient life—insects and plants provide the categories of being which provoke moral emotions and debate.

Now while this observation is not enough to *necessitate* the need for a personal lawgiver, the association of morality with the interpersonal and the sentient strongly implicates a provider with sensibilities attuned to self-other behavior and involving *mutual* concerns such as fairness, justice, harm, and so on.

In the case of Darwin noted above, his concerns about the morality of a god are clearly premised upon the implicit association of a "good god" with a "good person." No kind and thoughtful person, governed by the principle that one should not cause pain without good reason, would intentionally design a system the necessary workings of which entailed suffering. The "problem of evil," one might say, is *premised* on the assumption of a personal deity concerned to bring about the good in his creation.

Without the idea of a world created by God, we may acknowledge that bad things are to be found in it, but *without* the idea that the world has been created for the good, but simply physically exists without purpose, intentionality, or agency, the question of whether or not the world should contain bad loses any traction at all. Good and evil are just present in the way sand and rain are.

Without the presence of sentient intention, there can be no preceding question as to whether it was *intended* to be one or the other or both. The *problem* of evil only *arises* in the way it does for the theist because believers acknowledge that there is a prima facie conflict between the claim of a *personal* god acting for the good of those things that he brought into existence, and the presence of harm and suffering in his created order.[60] To acknowl-

interests which could be the subject of moral concern, is not clear. The theist, on the other hand, might argue we need to treasure the physical world because it belongs, as it were, to God, and is constructed from teleological principles of value. Alternatively, or at the same time, it may be argued that damage to the environment causes potential harm to come to sentient creatures, and *this* is a bad thing. In any event it will be clear that the central force of ethics derives from their person-regulating functions, or their sentient-creature-regulating functions.

60. See the section on the moral argument in Moreland et al., *Debating Christian Theism*, for examples of discussion in this area. See also Hunter and Nedelisky, *Science and the Good*. They trace a number of attempts to establish a naturalistic or scientific foundation for morality, including those associated with E. O. Wilson, Patricia Churchland, and Sam Harris. Concluding that these attempts have ultimately been unsuccessful, they argue that much contemporary "moral science" has shifted toward denying the objective existence of right and wrong, reconceiving morality instead as a project of socially engineering happiness. In their view, this amounts implicitly to a form of moral nihilism.

edge this is also, of course, to begin from the intuitive assumption that the person or persons of God are intimately involved not only with us but all the creatures and elements of the created world.

So, I conclude that the content of the classical arguments, while resting on "what can be known or rationally believed about God on the basis of human reason, without recourse to any special or supposedly supernatural revelation,"[61] also rest on the perception of some relation of God's person or persons *to* the natural that may be perceived *in* the natural.

Finding God, the Person, in Nature and Its Objects

That the naturalist, as well as his antagonist, the theist, both continue to experience nature, as expressing characteristics of the personal, prescinding from the experience only with cognitive effort, should suggest we take seriously that *nature* might, in fact, present us with such experiences because they *actually*[62] manifest them. If nature can manifest mindedness and agency in us, *might* the natural world express similar manifestations of the divine in its explanatory structures? At least, we should be prepared to consider that it might.[63]

In his work based on his St. Andrews Gifford Lectures of 2017, *The Hiddenness of God*, Michael C. Rea has a chapter titled "Divine Presence in a Material World."[64] His concerns differ from mine since he is interested in offering some insight into how *any* divine encounter is possible. He highlights the "traditional view, that *if* nature or the material world has properties pertinent to 'meaning' and 'value,' that are characteristic of the sentient life of persons, then they do not fall *within* the scope of the natural world as characterized by the sciences." However, as he writes, reflecting on "a broad strand of thought in contemporary philosophy and theology,"[65] there is a trend to

61. McGrath, *Fine-Tuned Universe*, 22.

62. I use the word "actually" in its strict Aristotelian sense to suggest that an "act" is a coming into being of a potentiality of a substance.

63. I note that we might here be talking about not only many different conceptions of "nature," but experiences of the personal *in* nature as differing in character, extent, and so on from different science to different science, and from culture to culture. Nevertheless, I claim that *all* human persons make some distinction between what is and is not them. Even in the case of an animistic society that detects the hands of spirits and gods in the natural, the nature of those spirits and gods will be such as to enable distinctions between them and the *Homo sapiens* who claim to detect or experience them.

64. Rea, "Divine Presence."

65. Rea, "Divine Presence," 128.

> buck against the modern conception of the world as "disenchanted" or entirely free of properties that fall outside the scope of what is investigated by the natural sciences.[66]

I have already drawn attention to this movement in previous chapters, and indeed this work is an attempted contribution in that direction.[67]

What is meant and entailed by this distinction is crucial. As I noted in the discussion earlier on Gregory Dawes, the presentation of the world to us as "natural"—i.e., devoid of the personal—is a pragmatic one that can exclude the necessary metaphysical structures that allow for its investigatable isolated "physical" appearance to us in this manner.

As I pointed out, the metaphysical background is taken for granted by the modern scientist. As I further argued, matters are not that simple. Theoretical evaluation often does include what might be classed as "normative" properties. As the reader will be aware, I have maintained that, in the realm of mathematical physics at least, nature displays properties that are precisely those of an aesthetically appealing character. Whatever else they are, they seem to fall within the scope of mathematical physics and are precisely characteristics, I suspect most would agree, that are essential to the perception of anything as "enchanted."

Now Rea's primary interest is in showing "how coming to be able to experience divine communication in the material world might require a certain skill that will be in principle learnable by anyone who has the concept of god."[68]

To be possible, however, the nature of things must be such as to allow for divine encounters in the context of the development of the appropriate skills. And here Rea agrees with me.

To introduce the view towards which he wants to move, he begins by highlighting a thesis he suggests is presupposed by all attempts to respond to the mind–body problem. He intends to include analogical reasoning, inductive reasoning, and Inference to the Best Explanation. If I have not misunderstood him, he argues that all these traditional attempts at solving the mind–body problem rest on the premise that the mind is "hidden" behind behavior.

66. Rea, "Divine Presence," 128.

67. Panchuk, review of Rea, *Hiddenness of God*, provides an interesting overall assessment of the book.

68. Rea, "Divine Presence," 129.

He goes on to argue that the problem of other minds rests on the idea that we have no "special *causal* contact with other people's minds."[69] He goes on to say,

> The problem of other minds is serious only on the assumption that we have no such contact, so that every physical event we are aware of is at least adequately explained . . . without appeal to other people's minds.[70]

His next paragraph contains the critical remark:

> The only way we manage to experience the presence of other minds within the material world is by way of cognitively impacted experiences whose stimuli are natural, physical phenomena. . . . Reflection on other minds points to the conclusion that we would experience God's presence in much the same way as we experience the presence of created persons.[71]

Rea's continuing argument is long but, for my purposes, may be adequately summed by his idea that it is in the recognition "that there exists another subject for whom I am the *object of perception*"[72] that constitutes my becoming aware of another person. What he wants from this is a model of how our perceptual confrontation with nature provokes a similar awareness of the divine personhood behind it.[73] He uses Sartre's example of feeling shame, which is intrinsically related to my responding to another as a person. To feel shame *is* to recognize that I am perceived by another.

This is an acute analysis of the *psychology* of the recognition of another. It may not, however, explain *how* matter which is not uniquely associated with the behavioral expressions of others successfully communicates to me personhood. To "feel shame" I must already *be* responding to the other as a person. My response cannot therefore be the *cause* of my acknowledging the personhood of the other, but a consequence *of* my acknowledgment that it *is* a person whom I observe.

The trouble is that although he acutely observes that my recognition of myself as a subject *entails* my recognition that I am experienced as a subject by another, it is not entirely clear that Rea is free of the traditional mind–body problem. What he has advanced is a *psychological* theory of how we

69. Rea, "Divine Presence," 131.

70. Rea, "Divine Presence," 132.

71. Rea, "Divine Presence," 132.

72. Rea, "Divine Presence," 132.

73. As I noted before, however, he hopes to extract a general account of what experience of God involves.

recognize each other once the epistemological channels to do so are open, *not* how our "restricted" perceptions of bodily movement allow for this possibility, at least in the mind–body dualism set up by Descartes.

He wants to rest his account of our encounter with God on a logical parity between the subjectivity of the interpersonal and our recognition of the person of God's subjectivity in nature. If he has not succeeded in the former, then while he may be entitled to a *psychological* account of our response to perceiving God in the universe *once* we perceive him in nature, he is not entitled to claim that his theory accounts *for* that perception.

How do we get to that perception? The whole idea of this thesis has been to argue that in recognizing design in the world through metaphysical spectacles, we approach a teleological explanation for its coming to be and for its structures. That we find some of them to be beautiful enhances our proclivity towards perceiving intention behind them and in them. Finding something beautiful can bring with it a strong sensation of the creator, what they might value, and a deep sense that creating real beauty is not simply a technical skill but the outcome of deeply felt values and appreciation for which the beautiful thing is a vehicle of expression: for an object to express these things is for it to have been "designed" in such a manner that the structure of the thing embodies that Beauty and what it entails.

Once our capacities for reacting, understanding, and engaging with an artifact as the product of something personal are mobilized, so are our sensory, emotive, and imaginative responses. In beauty we are, or can be, commonly transported to those things associated with it. "Beauty" is a good. Being a good, it enhances and promotes those values associated with it, such as love and affection and a desire to get closer to the beautiful thing.

Consider, then, the power of the acknowledgment of the absorption of the fact that a part of nature, the part of it at the heart of its origin and emergence, turns out to be "designed." If we take the full power of this seriously, we are in the position I outlined by D. C. Schindler:

> [Beauty has] an astonishingly comprehensive quality . . . it appeals to the whole of us . . . engages our mind and senses at once, enlisting them in the common project of perceiving beauty. . . . Beauty thus involves . . . our sense of transcendence . . . our sense of being elevated beyond ourselves . . . at the very same time it appeals to our flesh . . . our most basic natural instincts. In this way it reaches into the depths of our nature, touching what is most basic and essential.[74]

74. Schindler, *Love and the Postmodern Predicament*, 40.

We touch real personhood when we confront an object which we immediately take as an artifact because it communicates teleology to us. In so doing it can communicate something of the creator and his intentions. That experience is to become aware that a *real* personal life lies inside it.

To be aware of the teleological activities of God, as expressed in our world, through the structures of its creation is also to be reflectively aware that we ourselves are an outcome of the metaphysical possibilities of that world, and ipso facto of our own natural incarnation as an outcome of God's authorship. In this sense, a proper awareness of myself, via the presence of others, and our shared artifactual existence is already a personal engagement with God's artifactual world.

Chapter 6

Ultimate Answers

Meaningless, Impersonal, or Personal

SUMMARY

IN WHAT FOLLOWS I argue that any concept of the "natural" that does not include teleology and "personhood" cannot provide an ultimate explanation of what exists or why it exists or explain the beauty at its physical heart.

OVERVIEW OF THE CHAPTER

No conception of the origins of the world restricted to impersonal physical events, I argue, can provide an "ultimate" explanation for the fact of anything existing at all, its coming into being in aesthetic structural form, or, indeed, its having that form. I also contend that abstract entities, such as Platonic universals, or numbers, if they qualify as such things, cannot be sufficient for that purpose since they are causally inert.

For any concept of impersonal physical causation to be operative, a physically functioning "natural" world of structures must already be in existence. Nevertheless, such a world cannot be entirely devoid of the idea of creation. We are thinking creatures in that world. As agents therein, we constantly think and act in creative terms, asking how something might be designed, fashioned, or brought about in pursuit of particular ends. However incomplete our understanding of creation in its fullest sense must remain, the concept is therefore not without experiential content. It bears a partial analogy to our own purposive agency in bringing genuinely new things into the world and, in that respect, provides a conceptual bridge to the idea of an ultimate divine creator.

Recourse to the "personal," as well as offering an understanding of creative agency, also fulfills the metaphysical need to understand the normative and expressive aspects of the universe *including* its mathematico-physical character and the beauty of its ultimate constituents. The "divine" and personal ultimacy of all things coheres naturally with the case I have made for the "world" coming to be in varied structures, intelligibly accessible, and aesthetically organized to be functionally beautiful in its most basic physical forms: the intentional outcome of *artifex pulcherrimus*, the "most beautiful designer."

I devote a section of this chapter to discussing the nature of natural theology in the light of understanding ultimacy in the way just described above. I suggest that if natural theology begins from the premise of *accepting* the "personal" and "impersonal" metaphysical duality of the world that devolves from ontological naturalism, then it faces problems that are predetermined by that position: primarily a commitment to the primacy of the impersonal, including "abstract" reason and the "causally closed" physical world as providing the foundational certainties from which the task of proving God's existence must begin.

I look at one theistic example of this that suggests that "God" is a theory which explains all the "data" of existence that no "physical" theory could. The general critique of this position is that it assumes what the ontological naturalist assumes: that there is a metaphysical position "outside" of the world and God that human persons can occupy to determine the ultimate origin and nature of the world and ourselves in it.

In other words, the general position assumes that the objective stance available to the subject within the world to investigate its causal aspects is one that can be transferred to a position "outside" that world from which to theoretically investigate its ultimate origins. For the same reason I offered in the paragraph above, I suggest that there is no epistemologically neutral viewpoint from which to examine the proposition "Does God exist?" and I look at all the alternative ways in which the ontological naturalist might try to secure such a vantage point.

PART 1

Ultimate Reality

"Ultimate reality," Wesley Wildman begins a paper by writing,

> is the final word we can utter concerning reality as such in all its fullness, its general features held together with its concrete determinations. Such speaking is the heart of theology, which then radiates outwards to the vital tasks of interpreting nature, history, life, and humanity in the light of what is said about ultimate reality.[1]

Is it feasible to suppose that the "ultimate" origins of the cosmos could be incorporated into a final explanation of everything which excluded the personal and the teleological? Might the urge for ultimacy not rest, if not in purely physical elements, then in the finality of a developed naturalist metaphysics that can incorporate an explanation as to why no such recourse to teleology or the divine is necessary?[2]

Once the naturalist "structure" and all its ontological and epistemological relations to the sciences has been articulated, assuming that project could ever reach completion, would the finality of the understanding of everything be exhausted? Would our need for meaning and purpose be finally assuaged? Would the place and nature of personhood, teleology, structure, aesthetics, and value and free will finally become explanatorily transparent?

The puzzlement about ultimacy is not simply about the operations and function of physical items or, indeed, abstract items, such as the ontology of numbers, but embraces everything which may be reasonably supposed to underpin the existence of anything at all. It will be superordinate to the most general of questions which encompass metaphysics, science, subjectivity,

1. Wildman, "Really Ultimate Reality," 260.

2. Although I make use of philosophical arguments in what follows, I am treading a thin semantic line between invoking a personal deity as making sense of the world once perceived in things, and the idea of "explanation." My reason for this is that the very idea of things in the world being available for explanation is part of what is to be made sense of. Even in strictly human affairs the answer "It was done by a person" is, in one way, a complete teleological explanation; in another way, it has an inexhaustible semantic depth. The presence of novels, literature, history, poetry, painting, archaeology, psychiatry, and philosophy are all witnesses to this effect.

In the case of a divine personage, we should perhaps be even more wary of seeking "explanatory" finality. An approach that has been debated from a thoroughly philosophical point of view, in which the question of ultimacy is assumed to have a final explanation, is via the question "Is there a foundation for existence?" The answer to this when it is in the affirmative includes reference to the necessity of that foundation. I am assuming in what follows that whatever is taken to be ultimate, if it is ultimate, is necessarily necessary. If it is not, then I assume it cannot be ultimate.

In their book *Is God the Best Explanation of Things?*, Rasmussen and Leon explore this topic from both sides of the debate. Part 1 ("Is There a Foundation for the Universe?") presents a thorough philosophical exploration of many of the issues I discuss in what immediately follows.

and anything under the sun, including that venerable orb itself—in other words, questions of the ultimate nature and origin of everything.

This is a quest that has occupied human civilizations throughout the ages. In a book-length study titled *The Divine Quest, East and West: A Comparative Study of Ultimate Realities*, James L. Ford discusses a multiplicity of different versions, ranging over different cultures, times, and conceptions of supernatural or divine ultimacy.[3] Of course, answers to the question what ultimate reality is need not incorporate a divine element into their conception of that order. All, nevertheless, in some way or another, attempt to meet the sometimes-inchoate human need not just to know how things work but to answer the puzzling question as to why there are things at all and the meaning of their presence.

The Illusion of Ultimacy

One response to the questions above is to deny that the very *idea* of an ultimate reality is coherent, let alone ultimate reality itself. This is the position argued for by philosopher Stephen Maitzen. His argument is spelled out in a work titled, appropriately, "Against Ultimacy."[4]

He begins his chapter with the suggestion that

> much religion, as well as much philosophy of religion . . . concerns itself with what its practitioners take to be "ultimate" in one or more senses. . . . The kind of concern with [this] . . . characteristic of religion presupposes that the ultimate is *possible*.[5]

He analyzes the idea that there might exist what he calls an ontologically ultimate "concrete being"[6] that brings into existence all other forms of being, and concludes that the concept of "ultimacy" is not logically tenable.

He ends his discussion with the observation that "the admittedly seductive notion of an ultimate reality is an . . . illusion, something it's also time we humans outgrew."[7] Intellectual progress, particularly in the sciences, has

3. Ford, *Divine Quest.* His first chapter is devoted to outlining the many ways in which the term "ultimate reality" may be understood and the different domains of its apprehension. Lloyd, professor emeritus of ancient philosophy and science at Cambridge University, examines the cultural variations underpinning major questions concerning the nexus between human understanding and world in his illuminating study of different ancient societies: Lloyd, *Being, Humanity, and Understanding*.

4. Maitzen, "Against Ultimacy."

5. Maitzen, "Against Ultimacy," 48.

6. Maitzen, "Against Ultimacy," 51–53.

7. Maitzen, "Against Ultimacy," 59.

rendered all such questions obsolete, he maintains. He certainly thinks that contemporary metaphysics and logic are sufficient to rule out the idea *tout court*, offering philosophy as an alternative means of transcendence.

> Philosophy achieves . . . transcendence by digging below the superficial commerce of life and asking the most basic questions we can ask about reality.[8]

I discuss further on whether or not philosophical discussion is adequate to the kind of absolute refutation of theism that he espouses. It is curious, however, that he suggests philosophy can ask the most "basic questions we can ask about reality" but rules out the most basic question one *can* ask about reality as logically impossible. One cannot help the thought that the only kind of transcendence that he is talking about is one that is already shaped by a particular view of what the world is like.

He cannot escape the question, however, since his exclusion of an explanatory deity or separate "ontological being" simply implies another kind of ultimate answer: ultimately the nature of reality is that it is itself ultimate, it has no reason or cause. Neither does he seem concerned that he himself has assumed that there is an ultimate philosophical method of determining whether the question of ultimacy has any sense before the question itself is ever addressed.

Maitzen's assumption is that reason, formal logic, and certain metaphysical essentials about the "property of being" are sufficient for him to offer an "ultimate answer" to the question of the possibility of even postulating divine ultimacy: The analytic method provides the logical technology which can adjudicate the meaningful content or otherwise of the claim. The illusory content of the idea of ultimacy having been demonstrated by logical analysis, it must follow that the desire for a divine answer to the question is one that logically cannot be satisfied.

"Ultimate being" in Maitzen's conception is a strictly metaphysical concept, logically encompassable within any scheme that explains the world, despite its nature as an ontologically independent "concrete" being on which the existence of all other concrete beings is causally dependent.

Now there is certainly a way of hearing this with which the theist might have some sympathy, but the bloodless, isolated, causal object, akin to a natural "superpower," separated by logical necessity from the objects of its productive capacities is not the "ultimate being" of theism. "Ultimate being" in theist terms has a sense and meaning which cannot be confined to the

8. Maitzen, "Against Ultimacy," 60.

simple and rigid identity of content which Maitzen requires to adjudicate whether it has any logical coherency.

What I mean by that can be understood if we try to understand what the theistic desire for God represents. It is one that is felt from every aspect of one's being—rational, emotive, imaginative, spiritual, psychological, and even corporeally, as a visceral hungriness. All these features of what it is to be a human person form an inseparable unity with a unique emergent property of desire not reducible to any one of them and certainly not semantically divisible into the separate logical propositions required for traditional analysis. It is this desire that an experience of beauty at the heart of physical existence can motivate.

It is a desire one might describe as reflecting a question about the totality of being as it emerges self-consciously through one's own experience as participating in the existence of the world. It is a sui generis desire which can only be satisfied by a sui generis transcendental unity equal to the task: this is the ultimacy the theist aims at, not the pallid "concrete" stolidity of Maitzen's analytic reduction.

It is a philosophical presumption that the difficulty in extracting a semantic core for rigorous logical analysis, even the impossibility of that task, demonstrates the lack of any content to the desire or attempts to articulate it.[9] Maitzen's conception of the task he thinks he has to meet is too conceptually suffocating to grasp the ultimacy expressed by theistic desire.[10]

The sense of transcendence the theist wants straddles the boundaries of our language in ways which it may not be able to finally meet semantically.[11] Whether or not this is sufficient to condemn it in analytic terms

9. Wildman defends a view he calls a "radically apophatic conception of reality." "Really ultimate reality," he says, "whatever it is, is beyond being, and thus beyond the putative gods that are held to be beings." Wildman, "Really Ultimate Reality," abstract.

10. Wilczek, "When Words Fail." The Nobel physicist, in a brief article in the prestigious journal *Nature*, discusses the difficulties that even physicists working at the base level of mathematical physics have in developing, using, and finding words and concepts to encompass new thoughts, discoveries, and mathematical descriptions. It is perhaps worth quoting his quotation of the eminent twentieth-century physicist Hermann Weyl: "Weyl stated in his Philosophy of Mathematics and Natural Science (1949), '. . . only to the gaze of my consciousness . . . does a section of this world appear as a fleeting image in space which continuously changes in time'"; ellipses original. It is not clear why divine reference should be any less semantically simple than the language of mathematical physics.

11. White, *Talking About God*, provides a sophisticated treatment of the problem of language and speaking about God. In the book he examines Aristotle on analogy, Aquinas on attribution, and Kant on religious language. In his final chapter he states, "It is . . . clear that if we are to talk about God at all, the language we speak must be in some way appropriate to the task. That . . . implies the possibility of some sort of comparison

in which only contents decomposable into definitional form, suitable for symbolic representation and deduction, have any final legitimacy over our ultimate experience of what the world represents, is itself debatable. As John Cottingham has asserted,

> The philosophical temptation is always to classify the transcendent. If this means to give a positive characterisation in literal language, such an undertaking is impossible. . . . That which is dimly apprehended . . . is, as mainstream theology has always asserted, beyond human comprehension.[12]

Given the restrictions on semantic clarity when talking about the "ultimate," what we have to aim at is in the epistemological spirit of Wildman's observation about the possibility of rational knowledge of ultimacy derived from analysis of the natural world.

> Traction between ontology of nature and ultimacy metaphysics is . . . neither extremely strong nor entirely absent, but rather such as to have the effect of making some views of ultimacy less plausible and others more plausible.[13]

Naturalism and Ultimacy

Naturalism accepts in some of its versions that there *is* a question about ultimacy to answer. The answer is argued to end with the presence of the "natural" itself in some form, and sometimes with a metaphysics to accompany it. This answer, if it accepts a naturalist metaphysics at all, locates the origin of ultimacy in the structures of any metaphysics it espouses. Thought cannot proceed beyond those structures because from this perspective "beyond" can mean nothing.

Many atheist naturalists of considerable philosophical agility contend that the question of ultimacy is not one that has an answer, not because the answer remains mysterious but because the posing of the question itself already presupposes that a certain sense can be made of it. *That* they deny. It is an illegitimate extension of the ordinary sense of the word into an arena

between God and his activities and human beings and their activities." He supports the basic view of Barth and Aquinas that "the primary or proper meanings of the words we use . . . about God is in their application to God. . . . Their application to humans involves a secondary meaning." White, *Talking About God*, 187–88.

12. Cottingham, *In Search of the Soul*, 150.

13. Wildman, "Comparative Natural Theology," 174.

of metaphysical vacuity sharing a seductive syntactic structure with the mundane sense.

There are different versions that this might take. Maitzen's case may be thought as an example, although he concentrates on the logic of ultimacy rather than ruling the question out *tout court* on the grounds of empty content.

The most powerful of the naturalist arguments associate their position, not unsurprisingly, with the achievements of the sciences, or particularly physics, which they take as the epistemic gold standard for ontological decision-making. Rather than present the variety of these, I offer a particular example which appears as a chapter titled "Naturalism Without Metaphysics" by John Collins, a philosopher. Collins is interesting because he attempts to combine naturalism with an ontological empiricism while denying a commitment to ultimacy.

Collins maintains that

> on a standard view, naturalism is foremost an ontological or metaphysical doctrine about what there ultimately is as specified by some notion of "basic science."[14]

However, he resiles from claiming this position for himself while asserting his naturalist credentials. Here is how he motivates that claim:

> To be a naturalist on my favoured conception is simply to take one's lead from the relevant ongoing empirical enquiry. No metaphysical quandaries arise here or animate the naturalism save for those that arise in the interpretation of the relevant theories.[15]

He defends, he tells us,

> a version of methodological naturalism which stands against any brand of metaphysical naturalism worthy the name. . . . In my sense of the term naturalism is an anti-metaphysical doctrine.[16]

Collins, if I have understood him correctly, takes it that any ontological question can only be deferred to what is empirically extant in some domain of scientific practice. If empirical practice should have no end, then so be it. The nature of physical things would simply have this character.

14. Collins, "Naturalism Without Metaphysics," 85.
15. Collins, "Naturalism Without Metaphysics," 85.
16. Collins, "Naturalism Without Metaphysics," 85.

While there is much to be said about his thesis which he works out in detail, it is hard to see how he avoids the force of Nicholas Maxwell's observation which I gave in an earlier chapter:

> If science is to be possible at all, some kind of assumption must be made about that of which we are most ignorant: the ultimate nature of the universe. . . . *It is above all here that it is vital that we make as good an assumption as possible, one which does the best justice to the nature of things. For it is this basic cosmological assumption that will determine our methodology, what kinds of theory we are prepared to consider.*
>
> If, for example we believe a society of gods governs the natural world according to whim, then in seeking to improve knowledge and control over natural processes, it will be entirely rational (relative to this belief) to adopt such methods as . . . sacrifice, oracles, omens (and so on). If we believe a pattern of physical law governs natural processes, quite different methods will be adopted, namely those of putting forward precise hypotheses concerning possible laws governing phenomena to be tested against observation and experiment. Metaphysics determines methodology. . . . A bad basic metaphysical conjecture, hopelessly at odds with the basic nature of the universe, will lead to the adoption of an entirely inappropriate set of methods.[17]

If Maxwell is right, and I argued that he is, then the empirical world on which Collins relies will be dependent on the metaphysics from which method emerges, despite his denial that he is committed to any metaphysical position. Empirical conjectures involve the range of normative concepts I have discussed, the role of evidence, theoretical virtues that extend beyond measurement, acceptance of teleological demands that one "ought to aim at the theoretical truth," the understanding of the concept of "evidence" as a normative relation to the that "truth," and some accurate conception of the most general characteristics of existence.

Collins's position thus undermines itself. Firstly, it is itself a metaphysical picture of how things ultimately are: they are ultimately such that all that can be said to be "known" about "reality" is the current state of empirical play. In one sense, Collins is perfectly entitled to his metaphysical story, but in another, he cannot claim for it the same epistemological status as the empiricism he rests his case on. His convictions are not themselves empirically founded but a metaphysical *interpretation* of what empiricism means about the ultimate nature of things.

17. Maxwell, *Comprehensibility of the Universe*, 4–5.

The Normative Requirement of Evaluating Ultimacy

Let us even suppose that Collins is right about the future empirical state of play; there will be no possibility of sensibly developing an overall picture from naturalism or elsewhere, simply a never-ending following of the empirical trail. A question would still remain: Why should this be so? Why should it be that the empirical nature of the world is like *that*?

Whatever argument is presented by any version of naturalism, a similar question always remains open: Why should the world be that way? Whatever attempt the naturalist makes to close that gap in his own terms, the final resort must be that some "physical" or "natural" entity, *x*, or some physical entity plus some abstract, *y*, has the last word.

To have the last word, however, is always to step outside the physical and the empirical and make an evaluative judgment about what is to *count* as a completed picture of the ending of questions of intelligibility and ultimacy. That task requires the setting of criteria of "evaluation"—to wit, "When *x* is achieved, then questions of a certain metaphysical kind concerning ultimacy lose semantic traction." But it should be clear that this is an attempted definition which cannot derive from within the naturalist schema since there is no way of "standing outside it." Yet its presumption is to offer a view from outside of it in virtue of which "completion" could be successfully announced.

Even supposing the naturalist were right, he would still be engaged in a metaphysical question: that of limning the outer edges of reality. The question of ultimacy is still one which motivates this naturalism. It just places the *end* of ontology and being in the existence of natural things themselves and denies a divine and personal cause beyond them.

Is this a sufficient termination point for ultimate questions? One kind of answer if one is unhappy with this but does not wish to invoke personal agency is to propose, in the Platonic style, an abstract, unchanging infinite world, to which our world owes its genesis.

Nonphysical Causes and Ultimacy

In his discussion of the creation, which he subtitles "Event Causation or Personal Causation," Douglas Groothuis[18] mentions that D. Willard[19] and others have suggested that an argument might be made for the existence of a nonphysical, originating cause outside the physical universe. One may

18. Groothuis, "Metaphysical Implications."
19. Willard, "Three-Stage Argument."

deny the existence of God, or any divine personhood however conceived, but accept that the ultimate picture of the world requires abstract items to explain its existence.

Both Plato and Aristotle, for example, in their distinctly different ways thought that world and its abstract and concrete particulars entailed some final entity, but neither ascribed to that entity properties of human personhood. Plato wanted that entity to be the final concatenation of "good," "beauty," and "truth," but the form embodying these qualities would be fully abstract and nonpersonal in his vision. Aristotle wanted a "prime mover" to stop an infinite regress of changes, impersonal and passive, the necessary logical terminus of a sequence involving his four causal principles.[20]

The major criticism of positions that postulate unchanging impersonal entities as originating source for the world is that while such entities are eternal and unchanging, they lack causal powers. Groothuis gives an example from mathematics:

> The number seven taken platonically, causes nothing. . . .[It] may be instantiated in beliefs, utterances, or sentences, but . . . does not originate anything. Therefore, [abstract entities] . . . seem to make poor candidates for causal support of the entire universe.[21]

Infinite Existence and Ultimacy

Another kind of answer is that the world and everything ontologically necessary to it has no terminus outside of itself. Elements that emerge or disappear will ultimately be causally dependent on those whose existence in the world have an infinite guarantee. This modification gives ontological space to the evolution of natural events, wherever and in whatever shape they occur, including the emergence of sentient and rational creatures. The scientific and philosophical aim within this context will be to finally grasp the nature of the relations within the ultimate natural structure that explain the various emergent features of the universe as we now have it.[22]

The infinite existence of something, however, does not foreclose the question of its origin. We have to distinguish between causal explanations of

20. Carroll, "Aquinas and the Big Bang," offers an accessible summary of Aristotle and Plato and the classical Thomist response to the claim that an infinite universe disconfirms the theistic belief in creation. See also Aquinas, *On Creation*.

21. Groothuis, "Metaphysical Implications," 110.

22. See Rasmussen and Leon, *Is God the Best Explanation*, part 1.

things whose coherence involves temporal relations—*x* occurs before *y*, for example—and logical priority. How something came into infinite existence is a question of the latter form, not one involving temporal relations. From the logical point of view, it is a perfectly sensible question.

It cannot be answered by reference to any natural physical phenomena, however these are conceived, because any natural phenomenon is distinguished by its necessary relations to other natural phenomena in space-time. Since any physical object or event comes into existence and is ultimately dependent for its contingent status on the basic forces of particle physics (or any other postulated physical structure), no other physical object in the period following its coming into existence could be the cause of space-time.

Neither could space-time itself be the cause of its own existence. This is because it would have to have had the *potential* to come into existence. But nothing can have a potential before it exists. Potential implies that a change can occur from one thing to another: from having property *y*, for example, to losing property *y*, or vice versa. By definition, *before* there is space-time, there is no space-time, so there is nothing with any potential to bring about space-time.

This argument is not answered as, for example, the physicist Laurence Krauss[23] and others suppose it to be by the hypothesis that something *else* caused space-time: the emergence of space-time is then considered to be the emergence of the world as we have it out of "nothing." Krauss tells us that "nothing" is a concept that should now be left to the physicists. The objection to this is straightforward: to argue that "something" exists prior to space-time is not equivalent to the claim that "nothing" exists prior to space-time. Krauss is proposing to substitute "something" for "nothing" and claim an equivalence.

Krauss mistakes an existential category for an identifying one: in whatever way we are to describe or theorize about the antecedents of space-time, if it is to involve a proposition derived from physics, then it necessarily entails the assertion that "there is a least one thing (something) that exists." "Nothing" is the negation of that assertion: it has no referent and no content. "Nothing" and "something" are references to existential descriptors that are logically prior to *any* discussion and investigation of the world conceived under its physical aspects. The moment it is asserted that *x* precedes *y*, an implicit assertion is also made that *x* and *y* are something(s)—that is, that they exist.

23. Krauss, *Universe from Nothing.*

The Personal and Ultimacy

Groothuis's third option is the personal. Why, though, should the postulation of a personal creator make any additive difference to solving the problems raised above?

Groothuis answers by asking the reader to consider what he calls "event causation." By this he means that if creation is not a result of personal agency, then it must be the result of an "impersonal process or mechanism of some sort." This is the series of causal possibilities that I have been considering above. The trouble with this, he suggests, is that "examples of event causation always involve antecedent and contingent causal conditions, whereas *ex nihilo* creation does not."[24] I have asserted that just this is the case.

We take it, passé Hume, that explanatory relations between things should yield an outcome that makes it intelligible why *x* caused *y*. What makes it intelligible that the world as ultimately structured contains both subject and object, both natural and artifactual objects, is aesthetically decorated, and has a discoverable place for beauty in its physical heart?

Groothuis offers an answer:

> Instances of personal agency always grant the person some vital and determinative role in the actions it brings about. . . . Given our understanding of personal agents as unique initiators, viewing the (creating) being as a personal agent better comports with our experience of finite beings bringing about states of affairs through their personal understanding, will and power.[25]

If one accepts this thesis, then it inclines towards my position in something like the following way: Like artifacts, the world is an outcome of personal and transcendent creation. If not, then from whence are the origins of "beautiful structures" and "teleological artifacts" to be derived if that final existential resting place is taken to be impersonal? Of course, it cannot be immediately deduced from this question that divine personal agency must be ultimate, but it may have more plausible support.[26]

There are two reasons that contribute to the plausibility of this option. The first is based on the proposal that where two things have particular properties and events in common, then a relation between them is more likely than not. An appeal to the personal increases the intelligibility of

24. Groothuis, "Metaphysical Implications," 111.

25. Groothuis, "Metaphysical Implications," 111.

26. Rasmussen and Leon's *Is God the Best Explanation* offers a comprehensive argumentative exchange on this issue in part 2 ("Is the Foundation Personal?").

understanding the world because within the world are found personal life and agency and much else necessary for it, teleology, intelligible structures, normative standards, and beauty, for example, built into its fundamental mathematical structures.

The world considered on the other hand, as devoid ultimately of any such features in its essential origins and ultimately nothing but brute physicality, might be thought a lesser option. The question really is whether it is an option at all. If I have been correct in asserting the basic normative and teleological structure of things, particularly the aesthetic, as necessary for any contingent physical and natural world, then it is a logical impossibility that any natural world lacking these features could explain them, since its very existence depends on them.

While this is not on its own sufficient to invoke the personal, such an invocation better enhances the intelligibility of these properties, their presence and origins. Persons are intrinsically normative creatures, responsive to reasons, truth, and value. Theists have always held that these are properties shared by at least human creatures with God.

A perspective that includes the divine also participates in *an* understanding of how the world is organized such that it could be accessible to scientific intelligibility by creatures like us and offer ways of comprehending it under the rubrics "causal explanation," "teleology," and so on.

Secondly, the nature of our own experience contributes to the plausibility of an ultimate personal origin for the world. If *we* can create things that are separately beautiful from those that occur impersonally, and if beauty (or whatever aesthetic principle is pertinent to that part of the natural in question) cannot be explained by the purely physical or the impersonal, then whatever can explain the origins of our capacity to create beauty *and* beauty itself *composing* nature must be something in which both of these co-occur and is the origin of them together. The ontogenesis of beauty could not be derived *from* the basic causal and physical elements of nature because aesthetics contributes to *composing*[27] them as the physical elements they are.

If, in the end, wherever it is to be found, beauty requires the hand of the personal, do not these propositions taken together plausibly intimate a plausible defense of a creating designer who built aesthetics into the

27. I have already made the case that structure *and* aesthetics are metaphysically indissoluble. As to composition, I assume this principle of organization has a face-value logical integrity: if x is required to compose A, then, unless some complicated backward causation argument can be derived, it seems to follow that x has an ontological existence independent of that of A, but A does *not* have an existence independent of being composed by x. I put it that way because I do not want to claim that x is *temporally prior* to A. It seems clear that both could logically come into existence at the same time, but the ontological relation of A's dependency on x could still hold.

metaphysical fundamentals of the world and beauty into the physical fundaments of that world?

God as an Ultimate "Super Theory"

It is all too easy to elide the understanding of God as the ultimate "person" with a scientistic approach to what is meant by God as creator. In this section I concentrate on how the word "theory" can slip into discussions about the relations between God, world, metaphysics, and science, and bring with them a vulnerability to opening up the idea that "God" is a hypothesis.

This is the argument of Dawes which I presented in chapter 1: the hypothesis that God exists is an empirical "theory" to be tested against the evidence; it is simply a matter for methodological naturalism. Should the theistic theory succeed in explaining "everything," then the existence of God would be a matter of conceding to its explanatory empirical power: a "scientific" proof. Is this not what the theist should aim for?

This is not just a tendency that inhabits the world of the reductive physicalist and naturalist. It can find its way with equal facility into the theistic world. It is what has been called "religious scientism." David Alcalde, an astrophysicist and theologian, has recently dissected the consequences of what he suggests is a foundational conception of science as being free of metaphysical and theological presuppositions.[28] It is this foundational substrate which provides the illusion of a stand-alone world of physical causes and items which I have been concerned to contest.

His work has a different focus from mine, being largely concerned with a deeper exploration of the philosophy and history of philosophy of science that underwrites many positions in the understanding of the relation of God to science. Nevertheless, I essentially share the same general understanding of the fundamental misconception. His position is sympathetic to my expressed view that the picture of the relation of science to metaphysics is one of order and composition, although he is more concerned to trace out the consequences for theology than I have been. The back cover of Alcalde's book accurately summarizes his position:

> Despite its self-proclaimed neutrality regarding metaphysics and theology, modern science is full of metaphysical and theological presuppositions . . . a mechanised understanding of nature . . . [and] a reduction of God to an external agent in competition with natural processes, and creation to a worldly mechanism.

28. Alcalde, *Cosmology Without God?*

> . . . They remain unconscious for the most part in the dialogue between science and theology.[29]

A good example of how the delicate understanding of divine creation can collapse into the world of scientific understanding is contained in an exchange between the very distinguished Christian philosopher Richard Swinburne and A. Grünbaum, who achieved fame with his critique of psychoanalysis as a science.

The details of that exchange need not concern us here, but in his reply to Grünbaum, Swinburne writes this:

> When we come to theories of everything agreed to exist by theist and atheist alike (the Universe, its initial conditions, the laws of nature, mental life, etc.), there could be no scientific theories which make equally good predictions—it's theism or nothing. No scientific theory could predict that there would be a universe—that is, some physical objects—and that they would be governed by laws of nature; for scientific theories consist of purported laws of nature and they predict in virtue of the consequences thereof for initial states of physical objects, and so the theory could only restate and not predict its data. We need a theory of a quite different kind and of such theories my claim is that theism is the simplest and predicts the data well, sufficiently well to be made probable by the data.[30]

Swinburne's Argument from Theory: The Vice of Its Many Virtues

Swinburne is happy to assign to the divine hand the creation of everything science requires: physical objects, laws of nature, in fact "everything agreed to exist by theist and atheist alike." On the other hand, he takes it that theism is a theory that does a good job in predicting that everything that does exist, will exist.

The odd lacuna seems to be the missing of any idea of *creatio ex nihilo*. I say "seems" because the fault appears too obvious for such a distinguished theist philosopher to make. Nevertheless, if I am mistaken about Swinburne, it is certainly a commonly recognized feature of the debate about God and science. What is the error?

It lies in the failure to grasp that the very possibility of there being such things as "scientific theories" depend on the deep metaphysics allowing

29. Alcalde, *Cosmology Without God?*, back cover.

30. Swinburne, "Reply to Grünbaum," 481.

for "persons," "objects," "nature," and the relations among them. A "super theory" of all theories would itself be subject to metaphysical conditions.

So, the final understanding of "being" and "existence" cannot be a theory, since what a theory is and how the world facilitates the presence of subjectivity, objectivity, and epistemological access to its physical aspects are conditions of the word "theory" having a comprehensible content. In other words, the semantics of the word "theory" are confined to the epistemological relation between the personal and impersonal *within* the world mutually occupied by both.

Even if we suppose there is any sense in which there could be such a "meta theory," how could the truth or otherwise of such a theory be tested? Is there a mathematical description into which it could absorbed? How could the theory be experimentally examined? How is it possible to test Swinburne's "theory"? Did the theory "exist" before anything? Is the coming into existence of the universe a fulfillment of that preexistent theory? Did God hold that theory which he then put into practice? At this point the conditions of the "super theory" become incoherent.

What we are aiming for must be something that can supersede metaphysical conditions, something that transcends them. It must transcend the entire apparatus of world, nature, person, mathematics, logic, feelings, and so on ad infinitum. This is, after all, what "transcendence" means. So, the realization that God created everything, and God shares with us, however attenuated and however humble, an ability to comprehend that fact and see something of the world through his eyes, is not the outcome of a theoretical evaluation, whatever else it is.

What transcendence means in this context and what it carries us to is not incorporable within any one particular mode of understanding, interpretation, explanation, or causal theory available to us. It is sui generis. Is this a problem? No, it is precisely a sui generis terminus we are seeking, one perhaps that is the perfect sum of the above epistemological modes. This claim is akin to the traditional ascendancy to God through causation. The answer to the question of ultimacy once given as "God" allows all else because it is God who enables the various modes of comprehension of reality that we can aspire to, however humble and infinitely incomplete.

The best and closest we can come to understanding this is derived from our own capacities to create new items from nature. It cannot encompass *creatio ex nihilo*, but it participates *in* the *idea* of creation. The world God has given us is characterized, among its other metaphysical properties, by the intrinsic aestheticism of its structures, which are teleologically available to us, and whose properties we can exploit to create beautiful artifacts if not *ex nihilo*, then at least *de novo* with respect to the physical world that, in their beauty (if we wish), reflect its divine origins through their metaphysical heritage.

All too often the theist concedes that the human observer can occupy an ontological and epistemological viewpoint from the other side of the transcendental horizon. It is an impossible task formed within the fear and attraction of the scientific image. For the theist there is no metaphysical territory beyond God. The only position we can occupy is a vantage point provided for us within God's world of which we are part and from which we are permitted to extract some partial understanding of it as created and how it works.[31]

PART 2

Ultimacy, Natural Theology, Science, and Proof

What consequences does this have for natural theology? If the idea of an ultimate viewpoint from which to secure an argument for the existence of God from "natural premises" alone is not sustainable, how is natural theology to be understood? This question might be thought pertinent to my project since the presence of contingent aesthetic structures in science has formed a part of my case. The answer depends in part how the phrase "natural theology" is to be understood.

Alister McGrath in his 2009 Gifford Lectures says that the term "natural theology" has been widely used in Western philosophy since about 1750

> to designate what can be known or rationally believed about God on the basis of human reason, without recourse to any special or supposedly supernatural revelation.[32]

Christian theologian and philosopher John Frame, for example, describes the phrase as traditionally denoting the attempt to "learn about God through revelation outside the Bible."[33]

The term "natural theology," as de Cruz and de Smedt point out, can cover

31. Davies, *Why Beliefs Matter*, is a mathematician who argues that there is no "objectivist position," partly because he thinks that the absolutist conception of the objectivity of science is an illusion. He contends that a multilayered view of the universe is inevitable because it is constrained by human nature. While he might be sympathetic to my view concerning the richness involved in conceiving nature and the universe, he would not accept the unitary objectivist picture of the world entailed by my thesis.

32. McGrath, *Fine-Tuned Universe*, 22.

33. Frame, *Nature's Case for God*, 3.

> several fields of enquiry, including the formulation of natural theological arguments, the search for the natural knowledge of God that humans possess, and the conceptualisation of nature, including an interpretation of the natural sciences, within a theistic framework.[34]

"Natural theology" in the sense defined by C. Stephen Evans is the attempt to show that "there is a 'natural' knowledge of God's existence . . . that does not presuppose any special religious authority or revelation."[35]

He also notes that the term does not just encompass arguments for the existence of God but any attempts to deduce aspects of God's nature from what is available to human perception and reason alone, unaided by divinely inspired textual revelation.

Alister McGrath offers a category which he identifies as "*theologia naturalis*." His description incorporates the distinction between religious authority, revelation, and the natural but expands Evans's definition in a useful way. The term, he proposes, "denotes both physics *and* metaphysics—the philosophical exploration of nature, inferring the existence and at least some characteristics of the divine by contemplating . . . *the forces of nature*."[36]

Much will depend here on what is meant by "forces of nature." The task of natural theology, at least one which hopes to open the door to a rationally accessible personal deity, cannot be met if the existence of the "forces of nature" as stand-alone physical entities separate from the a priori conditions of their intelligible existence, as structures for example, is taken to be the foundation from which natural theology must do its work. The implication of this is the same as restricting the personal by definitional fiat.

To conceive of the "natural" as confined to the physical aspects of nature is to move within the conception of the world as a Cartesian "Machine," albeit a modern one in which the forces and elements of the natural world are not understood as isomorphic with "mechanical" devices like pistons and cranes.

One cannot *define* a "machine" as an impersonal device and then attempt to *extract* personhood, teleology, and rational consciousness from it. Descartes understood this and had to propose another form of existence, the "conscious mind," to accommodate personal reality. It is of course precisely this problem that I noted the neurobiologist Terrence Deacon recognizes in bringing teleological concepts to bear in explanations of sentient life and

34. De Cruz and de Smedt, *Natural History*, 111.

35. Evans, *Natural Signs*, 1.

36. McGrath, *Fine-Tuned Universe*, 25.

the problems associated with it that McDowell is concerned to undermine (see chapter 1).

Because we are *part* of nature, we cannot elevate a distinction that we can make between different kinds of objects inside the world, which are part of the world, then elevate one kind of object, the "impersonal," into a transcendent platform from which can serve as the basis for a "proof" of God's existential authority. This would be like trying to prove that artists exist by examining the molecular structures of the basic chemicals of paintings.

There can be no more proving of God's existence from isolating one aspect of material existence than there could be a deduction from the kinetic activity of water molecules to the existence of a kettle. Or, for that matter, from the physical forces exerted on the buttresses of a medieval cathedral to the existence of something called a "cathedral."

It would be a very different exercise to begin with the artist and then examine how he used and mixed chemicals and how he manipulated them to create the effects he wanted. Here we might indeed learn many things about him, what he knew, what he intended, how well he understood geometric principles of perception, his mastery of paint mixing, his perceptual acuity, and so on.

It is in this sense that our examination of the different physical disciplines of nature might perhaps help advance our appreciation of the divine nature. Because we, too, participate in the components of creation, create artifacts ourselves, and are also an outcome of creation, we are allowed a glimpse of that creative nature.

If one already accepts the picture of a personal god embracing the totality of things, then this task is different. It will involve the development of a Christian theology of nature. This is what de Cruz explains as a "conceptualisation of nature, including an interpretation of the natural sciences, within a theistic framework."[37]

The Accessibility of God in Nature

This raises the question why God is not as easily accessible in nature to us as the personhood of other human beings is via their corporeality: bodily appearance, movements, and speech. Our certainty of the personhood of others is immediate. While the theist might suppose he perceives with equal facility God's person in nature, others clearly have no such experience.

C. Stephen Evans makes a case for two principles which are highly relevant here. In a section of his book *Natural Signs and Knowledge of*

37. De Cruz and de Smedt, *Natural History*, 111.

God titled "Why Would God Employ Natural Signs to Make His Reality Known?,"[38] he asks the question "What would follow from God's existence if it were true?"[39] He postulates that "one thing we might expect . . . is that knowledge of god would be widely available and not difficult to gain." He calls this the "Wide Accessibility Principle."[40]

The second principle is derived from the thought that although knowledge of God would be widely available, it should also be easily resistible. He calls this, unsurprisingly, the "Easily Resistible Principle."[41] His justification is offered in the following quote:

> Though knowledge of God is widely available, it is not forced on humans. . . . To allow . . . people this option, it is necessary for God to make the evidence he provides for himself to be less than fully compelling. It might, for instance, be the kind of evidence that requires interpretation, and include enough ambiguity that it can be interpreted in more than one way.[42]

It is perhaps worth observing that as history brutally teaches us, even the recognition of the personhood of other human beings is not as obvious as it seems it should be.

The Purpose of Intended Proofs of God

Arguments for God's existence, whatever their origins in reason or nature, are, as Evans's argument implies, in any case rarely persuasive on their own account in persuading theists to become atheists because they find one or more of the arguments to be logically flawed, or atheists to become theists because they find the arguments logically compelling. They are a continuous source of unresolved dissension. As Stephen Evans observes,

> [they] are frequently criticised and declared to be comprehensively refuted, and yet . . . continue to be presented. Some people, including . . . philosophers and scientists . . . find the arguments convincing, others . . . equally well trained . . . find them to be without merit.[43]

38. Evans, *Natural Signs*, 12.
39. Evans, *Natural Signs*, 13.
40. Evans, *Natural Signs*, 14.
41. Evans, *Natural Signs*, 15.
42. Evans, *Natural Signs*, 15.
43. Evans, *Natural Signs*, 2.

There is no unity among those who suppose that the classical arguments, for example, can or cannot provide evidence or rational support for a deity. There are also debates about what the arguments achieve even if they are successful. So, for example, James Sennett points out that a common response to classical arguments is to argue that even if they are sound, the deity of their conclusions " is a far cry from the elaborate deity envisioned by traditional theism. . . . One does not need to postulate a full-blown omnipotent, omniscient, morally perfect creator, and sustainer of the universe in order to satisfy the requirements of (the given) argument (a first cause, a designer etc.)."[44]

Both theologians and atheists may denounce the very idea of a natural theology for entirely different reasons. Notoriously, or appropriately, the great Protestant theologian Karl Barth argued that knowledge of God was via Scripture and revelation alone in the divine person of Jesus Christ, and natural theology an abomination.[45] As McGrath interprets the substance of Barth's objection, natural theology can represent the "human desire to assert itself and take control over things," seen by Barth as an "attempt on the part of humanity to understand itself apart from and in isolation from revelation."[46]

Some atheists denounce natural theology because in their eyes it is a subject without a subject matter: arguments derived from the natural are empty of content and their logical analysis is pointless, an exercise in trying to prove by reason alone that very small teapots circle the earth, undetectable by vision, telescopes, or any scientific instrument. This analogy was originally proposed by the English philosopher Bertrand Russell (1872–1970) to support his claim that the burden of proof rested on those who made such assertions, not on those skeptical of them. Theistic propositions, he said, were of this kind.[47]

Some arguments might be more persuasive than others for different people. Thus, someone maybe persuaded that the moral argument, the idea

44. Sennett, "Hume's Stopper," 82.

45. Holder, *Natural Theology*, reexamines the question of Barth's relation to natural theology.

46. McGrath, *Fine-Tuned Universe*, 19. McGrath also points out that Barth's "critique was directed against a very specific understanding of an attempt to prove God's existence or gain knowledge of God under conditions of humanity's own choosing." There are also suggestions that Barth was motivated by the political and cultural circumstances of the early to mid-twentieth century, particularly the rise of fascism, to restrict knowledge of God to revelation and religious texts, thereby protecting theology from corruption by political ideology.

47. Van Inwagen, "Russell's China Teapot." This paper is among the many discussing the Russell example. Richard Dawkins uses a version of it in *The God Delusion*.

that without a god there can be no ontological grounding for morality, provides strong conviction for belief while inferences from the natural order of things, cosmological arguments, and the like are unpersuasive and trespassing on the territory of the sciences.

Asymmetrically, someone who wishes to believe might be persuaded that the world bears clear evidence of God's existence and find no empirical and logical contraindications to holding the belief. Yet again, another person might find any one argument persuasive but be unable to reconcile the presence of natural evil in the world with that of the concept of a good god.

These reflections prompt the question as to what purpose the classical arguments and natural theology serve. Are they simply about proof? Perhaps no philosophical positions have available to them watertight, irrefutable proofs.[48] If this is so, should it entail an abandonment of traditional arguments by the theist, a retreat to faith and revelation alone, or an abandonment of his theological convictions, or the dismissal of natural theology? Not at all.

It is hard to see how anyone might cleave to *any* philosophical, theological, or cultural position across a range of issues if rationally watertight proofs accepted by every protagonist in a debate were required. Should the absence, for example, of an irrefutable argument for the objectivity of morality require that one no longer make universal moral claims with any conviction? Does the lack of an agreed deductive solution to the problem of "other minds" engender severe doubts about those around and closest to me? Do I seriously doubt the presence of a world outside my own consciousness because of Descartes's methodological doubt that I may be being deceived by a demon that it is so? Similarly, does the absence of any agreed account of what criteria distinguish a science from a nonscience mean that the practice of science should cease?

In an interesting chapter titled "Quam Dilecta,"[49] which appears as part of a book in which distinguished thinkers discuss their own relation to faith and reason, Peter van Inwagen, an influential analytic philosopher who became a Christian, addresses this issue directly. I can do no better than offer a couple of quotes from him that are useful in elucidating the issue. Van Inwagen begins by saying,

> Philosophers do not agree on anything to speak of. And why not? How can it be that equally intelligent and well-trained philosophers can disagree about freedom of the will or nominalism or the covering-law model of scientific explanation when each

48. Van Inwagen, *Problem of Evil*, lecture 3.

49. Van Inwagen, "Quam Dilecta."

> is aware of all the arguments and distinctions . . . that the others are aware of? How can we philosophers regard ourselves as justified in believing anything under these conditions?[50]

He goes on to offer a specific example. "How," he somewhat plaintively asks, thinking about philosophical engagements he has had,

> can I believe . . . that unrealised possibilities are not physical objects . . . or that human beings are not 4 dimensional objects extending in time and space . . . when David Lewis [another philosopher] . . . of truly formidable intelligence . . . rejects the things I believe and understands every argument I could bring in their defence?[51]

Considering the various options that leaves him with, van Inwagen suggests that the question must have some good answer. If it does not, he opines, the only other alternative is skepticism about *every* possible answer. There are good reasons, though, for supposing that no one is a real-life skeptic about everything, even with respect to deep questions—perhaps especially with respect to such questions—so he suggests that

> the religious believer is often held to a widespread double standard in writings about the relation of religious belief to evidence and argument. This double standard consists in setting religious belief a standard it could not possibly pass, and in studiously ignoring the fact that none of our beliefs on any subject could possibly pass that test.[52]

If we assume with some charity that both naturalism and theism are in the same position here, the question becomes what we are to make of the attempted proofs, counter proofs, and arguments that surround God's existence. A useful way of placing these in an interesting context, if not specifying how the skeptical question raised by van Inwagen is to be resolved, comes from James Sennett.

Sennett draws a distinction between what he calls a *logical* evaluation of an argument and an *alethic* evaluation of an argument. To brutalize his position for the sake of brevity, it is a disjunction that draws attention to the difference between rigidly examining a given argument for logical consistency and considering the argument in the context of a variety of other

50. Van Inwagen, "Quam Dilecta," 42.
51. Van Inwagen, "Quam Dilecta," 42.
52. Van Inwagen, "Quam Dilecta," 42, for the full argument.

epistemological possibilities, including the elucidation of the broader goals towards which it contributes.

This distinction is not unfamiliar in the practice of science in which the result of any one given experiment often depends on the strength of other experimental work, theoretical considerations, and so on. In a previous chapter, examples were given of mathematical results whose predictions were only fulfilled later, sometimes after considerable initial skepticism.

Finality in the Proofs of God's Existence

Belief in God draws on many facets of human life and thought, from hope, love, need, scriptural familiarity, yearnings for personal and existential meaning, and perceptual and rational encounters with nature. It is doubtful whether, in the light of all these elements, deductive reason or nature could suffice to provide final answers, although undoubtedly providing powerful tools in the quest.

Whether someone comes to a belief in God has a complex etiology drawing on every facet of a person's experience, thought, imagination, reason, theological exposure, spiritual development, and history. The outcome of all these factors when they issue in a position of faith may be thought of as a sense of the "ordering" of all things in the light of an understanding of their created origin. At least, the question of "ultimacy" as I am attempting to argue here, is answered, for the theist, by the rational, psychological, and imaginative satisfaction of the comprehended order of things in virtue of divine teleology.

It is not clear in any event whether the classical arguments show easily discernible logical flaws or not, so whether they fail the test of logical evaluation as straightforwardly as has been sometimes held after Hume (1711–1776) is, to say the least, contentious. In the light of this, the idea that they fail to illuminate or contribute to the illumination of our understanding of God derived from the natural world as part of a wider alethic goal is premature.[53] If this is so, what should the theist hope from them?

53. A recent collection of essays on Alvin Plantinga, *The Plantinga Project: Two Dozen (or So) Arguments for God* (edited by Walls and Dougherty), contains examples of highly technical debate concerning such issues as the ontology and origin of sets and their members, and the implications of collections of things into sets for the existence of God. These discussions exemplify the point that determining the purely logical status of the validity and soundness of such issues in relation to God is highly complex, if determinable at all. In Sennett and Groothuis, *In Defense of Natural Theology*, a series of essays contests the assumption that the Humean criticisms of the classical arguments possess the refutatory finality they are often taken to have.

It is worth repeating that the question of the role that logical arguments for God can play, which are derived from the perception of natural properties, from a priori reflection on them, or from scientific evidence, is more subtle than is first apparent. If God is the creator of *all* that exists, and all and everything that *could* possibly exist, then God is also the creator of the possibility *that* natural knowledge of his existence is cognitively available to creatures such as ourselves. It perhaps an underpinning worry of the absolutist appeal of ultimate deductive proof that the classical arguments might convey that concerned Barth—the idea that human beings can stand apart from God and exercise themselves a godlike capacity for reason.

The quest for natural arguments in support of ultimate existential questions and dilemmas should, as I have noted, perhaps be more sympathetically located as part of a broader alethic desire shared by all persons to understand and find meaning or purpose in the world in which they find themselves.

The natural theologian, understood as a part of this exercise, will be motivated by a deep desire to find in the satisfaction of an argument a rational resting place for his desire for God and a unification of his belief with other understandings of the world, including the empirical findings of the sciences. Saint Augustine's famous prayer, "You have made us for yourself, and our hearts are restless until they find their rest in you,"[54] beautifully expresses the final desire which motivates the theist.

C. S. Lewis's contention that humans have a natural desire for God because there exists something that will satisfy that desire, which is why it is the desire it is,[55] articulates an argument which develops Augustine's heartfelt thought. The classical arguments contribute to excavating part of what the theist means when he talks about "God" and "world," and in the course of doing so give more shape to what it is that he desires.[56]

The various natural arguments for his existence, conceived in these lights, are less proofs of his existence than akin to searches for articulating the ways in which reason, perception, and existence are intertwined with the encounter with his presence in the Cosmos. Their rigorous analysis is not, therefore, to be deprecated but applauded. The naturalist will embrace a different starting point. How, then, is he likely to conceive of classical arguments for God's existence?

54. See Augustine's *Confessions* 1.1.

55. Williams, *C. S. Lewis*, contains the full argument and a contemporary analysis.

56. Ellis, *New Models of Religious Understanding*, as the title suggests, is concerned with questions such as how to distinguish religious from scientific understanding, religious practice, and related issues.

Naturalists, Ultimacy, and Proofs of God's Existence

All naturalists hold that the nature of anything and everything is ultimately devoid of the divine personal in any form. They are necessarily atheists. Naturalists are at sea in a different existential boat from theists and will resist *all* the classical arguments. They might *attempt* to approach them with a fair-minded, objective vigor on the grounds of epistemological principle. One might suspect, however, that their efforts will be directed towards an effort to provide watertight refutations. If van Inwagen's contention accurately discerns the boundaries of what can be attained via pure logical endeavor, the naturalist will necessarily locate his analysis within a broader alethic struggle, that of showing that God does not, indeed *cannot*, exist.

This is an entirely honourable endeavor and one which should be welcomed by theists, promising to show them where more work needs to be done. It might even be contended that pushing naturalist arguments to the very limits of their existential application is a necessary contribution to the theological exercise since it can contribute to the elucidative potential of natural theology to our understanding of God and world.

It can sometimes seem that it is the theist that is on the back foot with respect to the classical arguments, but it is more rarely acknowledged that this rests on a presumption that they have been comprehensively refuted[57]—a conclusion which itself is, to say the least, contestable. The atheist and the naturalist have their own motivations: a desire that God not be found, and even a fear that he *will* be discovered.

A clear statement of this position comes from the American philosopher Thomas Nagel. He writes,

> I speak from experience, being strongly subject to this fear myself: I want atheism to be true and am made uneasy by the fact that some of the most intelligent and well-informed people I know are religious believers. It is not just that I don't believe in God and, naturally, hope that I'm right in my belief. It is that I hope there is no God! I do not want there to be a God; I do not

57. The widespread view that the classical arguments have been refuted appears to rest upon the assumption that David Hume comprehensively refuted them, and that Kant demonstrated the limitations of reason, including the impossibility of deducing God's existence from natural premises. The collection of essays previously cited, Sennett and Groothuis, *In Defense of Natural Theology*, presents several arguments on Humean themes which, in turn, offer extensive critiques of Humean-style refutations. The volume also contains discussion of the historical development of the perception of refutation. See also Conway, *Rediscovery of Wisdom*, for a refutation of Kant's argument. Feser's *Five Proofs of the Existence of God* similarly provides robust defenses of a number of arguments for God's existence.

> want the universe to be like that. . . . It is just as irrational to be influenced in one's beliefs by the hope that God does not exist as by the hope that God does exist.[58]

Nagel places the believer and the nonbeliever in the same queasy epistemological hinterland. In his book *The Last Word*, from which the above quote is taken, he acknowledges that the naturalist faces a problem of "cosmic authority" which he conceives of as a need for something like the divine for any sort of knowledge to be possible. "My guess," he writes immediately after the quote above, "is that this cosmic authority problem is not a rare condition and is responsible for much of the scientism and reductionism of our time."[59]

This is a reluctant acknowledgment of the naturalist atheist's predicament of rationality operating within the bounds of naturalism to answer every question surrounding the nature and origin of the natural. We cannot, as it were, take up "the view from nowhere"[60] without taking up a view from somewhere.

Many atheists rather differently might envisage a position that they *can* take "above the fray," a neutral viewpoint from which can be employed the tools of reason and evidence to adjudicate the classical arguments.

As I have already argued in a previous chapter, however, the atheist, when considering what ultimately exists, is, whether he likes it or not, *equally* committed to an a priori vantage point from which to view the Universe. He assumes that the universe offers an objective viewpoint free of any conceptual content from which God's existence *can* be considered or one whose content can be designated as epistemologically neutral.

This is not so. Logicians typically make a distinction between the validity of an argument and its soundness. One might display the syntactic form of any argument: for example, "If everything that is an x is a y, then if A is an x, A is a y." This a *valid form* of an argument and is a *well-formed theorem* of predicate logic. In *this* sense it might be called objective.

It cannot be assessed for *soundness*, however, since it has no content. The provision of an interpretation in the form of assignable appropriate content to the symbols provides for an assessment of soundness. Thus, "If everything that is an ox is also a fish, then if Bob is an ox, Bob is a fish" is *valid*, but not *sound*. One cannot thus reason from syntactic form to a conclusion about ontology, metaphysics, or God without beginning from

58. Nagel, *Last Word*, 130–31.

59. Nagel, *Last Word*, 131.

60. This phrase is drawn from the title of Nagel's seminal work *The View from Nowhere*.

a proposition or propositions that have *some* content. The question is from whence the content is drawn.

The naturalist might suppose that there is a standing point he can occupy which both contains content but is "neutral" with respect to that content. It should be clear, however, that that content cannot be neutral. It assumes that the world is able to provide a neutral viewpoint from within itself to examine itself from the outside. The conception of this possibility is incoherent.

The idea of a viewpoint bereft of content derives what meaning it has from ontological naturalism. That is, it assumes that the objectivity we can take up *within* the world to aspects of it can metaphysically expand to a perspective *outside* of the world. The theist might object that it *assumes* there is a point of view from which the very nature of being may be examined *outside* God's world, including his creation and God's self. This is *part* of what is at issue though, not independent of it.

The explanatory possibility that makes the naturalist's position seductive is that he *begins* with the assumption that Cosmic Authority is derivable from the physicalist universe out of which the conditions of its own existence or coming into existence can be extracted. *Within* the naturalist framework he believes he can find an objective stance which contains within itself the authority to answer questions about ultimate being. The achievements of the sciences bolster his confidence in this supposition.

These assumptions represent a confusion between the objectivity required for the sciences *of* the natural world and what is required to understand something of the coming-to-be-of the existence *of* a world in which it is possible to *take up* a position of knowledge-yielding objectivity at all.

The problem of the existence of a world structured so that one part or parts *can* be characterized as having a "nature" able to be empirically and objectively discerned is, I have already argued, metaphysical and theological.[61] The objectivity required for the natural sciences is a discovery by conscious reasoning creatures *within* the universe as they find it of how to reason and explore part of it. While that discovery might offer assurance about knowledge of things within the universe gained via those validated methods, it is no guarantee that they have cosmological authority over the final discernment of the ultimate origins of the universe.

61. As I argued in an earlier chapter, the modern conception of nature as a value-free field of physical objects possessing properties suitable for measurement and devoid of teleological properties is an achievement of modern science, philosophical reflection, and theology, rather than simply a "given" of perceptual or rational experience. Issues pertinent to this development are discussed in a series of essays in Harrison and Roberts, *Science Without God?*

The upshot of this conclusion is that the naturalist cannot claim a "natural" advantage over the theist when considering the view from which the classical arguments are to be assessed.

The naturalist might still insist that the propositions "God does not exist" and "God does exist" cannot both be true even if he concedes the above. If these are both propositions each of which has a claim to be true, then there must be something or nothing answerable to fill the referent of "God."

This being so, a resolution of the truth or otherwise of these contradictory propositions should offer a position from which reason can be used to adjudicate the truth. Therefore, he will claim that whatever his own belief, logic and reason provide an objective position of "no position" from which the truth or falsity of each proposition and its accompanying supportive arguments can be assessed. The trouble with this is that "not having a position" is precisely to assume that the nature of things is such that it is possible "not to have a position."

Now this might well be a useful and comprehensible strategy in all kinds of situations from the personal to the cultural to the scientific to the political. If what we are looking for, however, is a vantage point "outside" the world, then this position is not neutral with respect to content. It says that "the nature of things is such that they provide a position outside of themselves from which they can be examined." This is, however, the conclusion that the naturalist wants to reach, but it cannot be assumed to be true to begin with.

Another tack he might then try is to argue that the belief that God does not exist has the same claim to be "basic" as Plantinga's offering that God *does* exist is "basic." That all that there is is the natural, is as basic a place from which to start as the supernatural premise. This places him in an awkward situation. He demands that reason be the final judge, but since he, like his theist opponent, begins from a position which he accepts is *prior* to any reasoned warrant, he must be convicted by his own lights of unreasonable faith.

The atheist who has all his eggs in the ontological naturalist basket will most likely wish to avoid this assertion and argue that the proposition that God does not exist is a *conclusion*, warranted by the exercise of reason and evidence.[62] In the contemporary world he thinks that these tools are wielded by the sciences to yield the ineluctable conclusion that there is no god. This is the strategy pursued by Gregory Dawes discussed in an earlier chapter.

62. An atheist might not, of course, deploy the idea of "science," but his arguments will nevertheless suppose it to be true that a clear-minded examination of our day-to-day empirical world will yield the same conclusion. Lucretius's famous work *De rerum natura* (*On the Nature of Things*) has this structure.

Here the difference between the theist and atheist's epistemological boats are clearly marked. The theist is happy to acknowledge that he finds himself afloat in a sea, aspects of which might permanently—indeed are likely to—exceed his cognitive grasp. He can only lunge at them as best he can.

The atheist naturalist cannot afford such an acknowledgment. His premises rest on an assumption of evidential and logical supremacy whose illuminating lights can stretch into the deepest recesses of existence and its accounting.

He must explain how they have the decisive relation to ultimate questions of existence and being that he says they have. This task requires him to show how they fit into a metaphysical scheme of things that provides the warrants of supremacy he claims for them. That scheme cannot be a *product* of science since it must explain the *possibility* of scientific truth and the world it investigates. This demand by itself is not disarming. He might happily admit that the question is a philosophical and not a scientific one.

He will, of course, have to engage in an epistemological defense of how knowledge outside scientific endeavor is possible and widen the scope of his ontological claims. Since he denies any appeal to a Creator as a warrant for the truth-fulfilling efficacy of his epistemological tools, his options will be limited to whatever he argues constitutes the Natural.

This places him squarely, but not conclusively,[63] in danger of irrational circularity. Since he will only allow *certain* kinds of entities, relations, and causes to enter the sphere of his supporting metaphysical arguments, those he deems properly "natural," he ensures the outcome which he aims to prove.

This position will eventually *necessitate* points at which some beliefs become basic, since premises and axioms must rest somewhere, and where they do will have to be in the realm of the "natural."[64] However this is to be understood, it will have to exclude the "supernatural" understood minimally as an intentional actor external to the broadest characterization of the "physical" and what is needed for it available to the naturalist.

63. These questions have been addressed by a variety of thinkers with naturalist sympathies. Papineau defends the view that reason is part of the natural world and addresses several associated issues in *The Roots of Reason*. Reppert, in *C. S. Lewis's Dangerous Idea*, argued that if naturalism were true, then skepticism concerning scientific reasoning itself would follow. Quine is the classic locus for naturalized epistemology and the view that the only knowledge available to us arises from our sensory confrontation with the world, together with the mathematics and logic required for theory construction.

64. This does not depend on a "linear" series which terminates in a physical item. Were the naturalist to offer a picture in which there was no terminus but a "circular" structure in which everything depended on everything else, the question of the existence of the circle itself would arise.

This resting place will be a matter of intuition and conviction from which his reasoning begins. Such a starting point is one from which reason is *exercised*, not one which is *warranted* by reason. Classical logic is founded upon the principles of identity and noncontradiction: A = A, and –(–A and A). There are no proofs of them as a consequence; they are the foundations of all logical theorems and reason.

The naturalist thus finally stands in the same position as the theist as far as the classical arguments go. He must not only consider them from the narrow view of logical consistency but must consider their place in the overall alethic goals they serve. From this perspective he is likely to never rest easy with any conclusions drawn from the classical arguments and will be concerned to uncover the logical flaws he supposes must be there. Since his preliminary commitment to naturalist premises is a matter of nonrational conviction, it might be predicted that the resolution of the classical arguments will never find satiation as a matter of formal logic.

A Conclusion?

This outcome is not surprising. If, as I have suggested, faith and belief in God are the outcome of a growing relation and experience of God through the world, then no argument for his existence one way or the other could be indifferently rationally persuasive, any more than reason could successfully induce in me the conviction that my loved ones were bereft of consciousness, thought, and feeling because no comprehensive refutation or proof of the Cartesian split between mind and matter was available. Should someone make one or another such claim, it would, one might think, be inhuman and impersonal to ground my love and affection on that proof, or deny it, in preference to my certainty that whom I have come to love *is* a person.

A theistic view on the origins of atheism suggests with some nuance that a naturalist metaphysics absenting the divine has surprising theistic roots. This point is made with clarity by Thomas Broman in his chapter "Matter, Force and the Christian Worldview in the Enlightenment."[65] Broman asks us to consider that the "disenchantment of nature" associated with secularism and the rise of the empirical sciences in Europe, circa the eighteenth to nineteenth centuries, did not "suddenly begin" in the eighteenth century.[66]

In fact he notes,

65. Broman, "Matter, Force," 87.

66. Broman, "Matter, Force," 87.

> Its origin might be found as far back as . . . the moment when . . . God gave Adam dominion over the earth. A world subject to human dominion was not just . . . to be venerated as Gods dominion, but also to be understood and turned to human use.[67]

Broman then goes on to observe that

> enchantment and disenchantment . . . appear to have been built into our perceptions of the world . . . from the moment our remote ancestors began reflecting about their place in it.[68]

It is, however, remarkable that that reflection reveals a world necessarily filled with structures characterized by aesthetic features that *can* be perceived as ultimately personal in origin. It is strange that within the structures that constitute its fundamental physical base, exactly where the naturalist is most certain no enchantment is to be found, we find beauty serving not only rational purposes but glowing with beauty for its own sake.

If we add to this list of the extraordinary features of our world the observation that those beautiful structures at the base of all of what is physically describable also seem to obey the commands of logical and mathematical order, and that those mathematical structures have, at the very least, a profound association with our phenomenal experience of beauty, then the coincidences multiply close to impossibility. One might wish to call this a "metaphysical fine-tuning."

If we also acknowledge the extraordinary effects of beautiful things to move us beyond themselves, and consider how our own creations, when they move us in a similar fashion, palpably display the human heart which gave rise to them in their compositional form, it is not unreasonable that that coincidence should move us to perceiving our world, and us in it, as the outcome of a personhood greater than we can fully conceive but whose personhood is on display in the artifact of his profound creation.

Perhaps, in the end, if the hand of the personal cannot be experienced by someone through the beauty that shines through the basic structures of the world, there is no argument or evidence that could persuade them. Like the psychologist Edwin Boring's (1886–1968) use of the famous ambiguous illustration that can be perceived as either a young or an old woman without a physical change to the drawing,[69] seeing the world as God's artifact or not is not finally a matter for reason, but it is not irrational.

67. Broman, "Matter, Force," 87.

68. Broman, "Matter, Force," 87.

69. Boring, "New Ambiguous Figure."

Edwin Boring's ambiguous figure[70]

70. Illustration by W. E. Hill, "My Wife and My Mother-in-Law," 1915.

Bibliography

Alcalde, David. *Cosmology Without God? The Problematic Theology Inherent in Modern Cosmology*. Eugene, OR: Cascade, 2019.

Allen, Colin, and Jacob Neal. "Teleological Notions in Biology." From *The Stanford Encyclopedia of Philosophy* (Spring 2020 ed.), edited by Edward N. Zalta. https://plato.stanford.edu/archives/spr2020/entries/teleology-biology.

Almeder, Robert. *Harmless Naturalism: The Limits of the Sciences and the Nature of Philosophy*. Chicago: Open Court, 1998.

Ammereller, Erich. "Wittgenstein on Intentionality." In Glock, *Wittgenstein*, 59–94.

Aquinas, Thomas. *On Creation: Quaestiones Disputatae de Potentia Dei, Q.3*. Washington, DC: Catholic University of America Press, 2011.

Aristotle. *Aristotle's Physics*. Translated by C. D. C. Reeve. Indianapolis: Hackett, 2018.

Arrington, Robert L. "Following a Rule." In Glock, *Wittgenstein*, 119–37.

Ashworth, William B. "Christianity and the Mechanistic Universe." In *When Science and Christianity Meet*, edited by David C. Lindberg and Ronald L. Numbers, 61–84. Chicago: University of Chicago Press, 2003.

Augros, Michael. *Who Designed the Designer? A Rediscovered Path to God's Existence*. San Francisco: Ignatius, 2015.

Augustine of Hippo. *Sermons: 51–94*. Edited by John E. Rotelle. Translated by Edmund Hill. Vol. 3 of *The Works of Saint Augustine: A Translation for the 21st Century*. New York: New City Press, 1991.

Ayala, Francisco J. "Darwin and the Teleology of Nature." In *Science and Religion in Search of Cosmic Purpose*, edited by John F. Haught, 18–41. Washington, DC: Georgetown University Press, 2001.

Baker, Lynne Rudder. *Saving Belief: A Critique of Physicalism*. Princeton: Princeton University Press, 1987.

———. "The Shrinking Difference Between Artifacts and Natural Objects." *American Philosophical Association Newsletter on Philosophy and Computers* 7 (2008). https://people.umass.edu/lrb/files/bak08shrM.pdf.

Barbour, Ian G. *Myths, Models, and Paradigms*. London: SCM, 1974.

Barr, Stephen M. "Chance, by Design: The Scientific Concept of Randomness Is Consistent with Divine Design." *First Things*, Dec. 1, 2012. https://www.firstthings.com/article/2012/12/chance-by-design.

———. "The End of Intelligent Design." *First Things*, Feb. 9, 2010. https://www.firstthings.com/web-exclusives/2010/02/the-end-of-intelligent-design.

———. *Modern Physics and Ancient Faith*. Notre Dame: University of Notre Dame Press, 2003.

———. *A Student's Guide to Natural Science*. Wilmington, DE: Intercollegiate Studies Institute, 2017.

Barrett, Paul. *Science and Theology Since Copernicus: The Search for Understanding*. Pretoria: UNISA, 2000.

Benson, John, et al., eds. *Frank Sibley: Approaches to Aesthetics: Collected Papers on Philosophical Aesthetics*. Oxford: Clarendon, 2001.

Bernier, Rachel B. "After Aquinas: Restoring Hope to Beauty." *Philosophy and Theology* 28 (2016) 91–100.

Bernstein, J. M. "Re-Enchanting Nature." In *Reading McDowell on Mind and World*, edited by Nicholas H. Smith, ch. 12. London: Routledge, 2002.

Berry, Michael. "Paul Dirac: The Purest Soul in Physics." *Physics World* 11 (1998) 1–7.

Bloom, Paul. "Religion Is Natural." *Developmental Science* 10 (2007) 147–51.

———. "Religious Belief as an Evolutionary Accident." In *The Believing Primate: Scientific, Philosophical and Theological Reflections on the Origin of Religion*, edited by Jeffrey Schloss and Michael J. Murray, 118–27. Oxford: Oxford University Press, 2010.

Boring, Edwin G. "A New Ambiguous Figure." *American Journal of Psychology* 42 (1930) 444–45.

Boudry, Maarten, et al. "How Not to Attack Intelligent Design Creationism: Philosophical Misconceptions About Methodological Naturalism." *Foundations of Science* 15 (2010) 227–44.

Boyle, Robert. *A Disquisition About the Final Causes of Natural Things Wherein It Is Inquir'd, Whether, and (If at All) with What Cautions, a Naturalist Should Admit Them?* London: John Taylor, 1688. EEBO editions, ProQuest.

Breitenbach, Angela. "Beauty in Proofs: Kant on Aesthetics in Mathematics." *European Journal of Philosophy* 23 (2013) 955–77.

Breitenbach, Angela, and Daniela Rizza. "Introduction to Special Issue: Aesthetics in Mathematics." *Philosophia Mathematica* 26 (2018) 153–60.

Broman, Thomas H. "Matter, Force, and the Christian Worldview in the Enlightenment." In *When Science and Christianity Meet*, edited by David C. Lindberg and Ronald L. Numbers, 85–109. Chicago: University of Chicago Press, 2003.

Brooke, John Hedley. *Science and Religion: Some Historical Perspectives*. Cambridge: Cambridge University Press, 2014.

Brooke, John Hedley, and Geoffrey Cantor, eds. *Reconstructing Nature: The Engagement of Science and Religion*. Oxford: Oxford University Press, 1998.

Brown, Frank Burch. "Aesthetics and the Arts in Relation to Natural Theology." In *The Oxford Handbook of Natural Theology*, edited by John Hedley Brooke et al., 611–29. Oxford: Oxford University Press, 2013.

Bubner, Rüdiger. "Bildung and Second Nature." In *Reading McDowell on Mind and World*, edited by Nicholas H. Smith, ch. 11. London: Routledge, 2002.

Budd, Malcolm. *The Aesthetic Appreciation of Nature*. Oxford: Oxford University Press, 2002.

———. *Aesthetic Essays*. Oxford: Oxford University Press, 2008.

———. Review of *Real Beauty*, by Eddy Zemach. *Philosophical Review* 108 (1999) 100–103.

Bueno, Otávio, et al., eds. *Thinking About Science, Reflecting on Art*. London: Routledge, 2018.

Burtt, Edwin A. *The Metaphysical Foundations of Modern Physical Science*. London: Kegan Paul, 1925.

Carroll, William E. "Aquinas and the Big Bang." *First Things*, Nov. 1, 1999. https://firstthings.com/aquinas-and-the-big-bang.

Chandrasekhar, Subrahmanyan. "Beauty and the Quest for Beauty in Science." *Physics Today*, July 1979. https://physicstoday.aip.org/features/beauty-and-the-quest-for-beauty-in-science.

———. *Truth and Beauty: Aesthetics and Motivations in Science*. Chicago: University of Chicago Press, 1987.

Clark, Stephen R. L. *Plotinus: Myth, Metaphor, and Philosophical Practice*. Chicago: University of Chicago Press, 2016.

Collingwood, R. G. *The Idea of Nature*. Oxford: Oxford University Press, 1945. Reprint, Mansfield Centre, CT: Martino Publishing, 2014.

Collins, John. "Naturalism Without Metaphysics." In *Experimental Philosophy, Rationalism, and Naturalism*, edited by Eugen Fischer and John Collins, 85–109. London: Routledge, 2015.

Collins, Robin. "The Argument from Physical Constants: The Fine-Tuning for Discovery." In *The Plantinga Project: Two Dozen (or So) Arguments for God*, edited by Jerry L. Walls and Trent Dougherty, 89–107. Oxford: Oxford University Press, 2018.

———. "The Fine-Tuning Evidence Is Convincing." In *Debating Christian Theism*, edited by J. P. Moreland et al., 35–46. Oxford: Oxford University Press, 2013.

Conway, David. *The Rediscovery of Wisdom: From Here to Antiquity in Search of Sophia*. London: Palgrave Macmillan, 2000.

Cooper, Lynn A., and Roger N. Shepard. "Chronometric Studies of the Rotation of Mental Images." In *Visual Information Processing: Proceedings from the Eighth Annual Carnegie Symposium on Cognition*, edited by William G. Chase, 75–176. New York: Academic, 1973. https://doi.org/10.1016/B978-0-12-170150-5.50009-3.

Copan, Paul, and Christopher L. Reese, eds. *Three Views on Christianity and Science*. Nashville: Harper Christian Resources, 2020.

Corby Hovis, R., and Helge Kragh. "P. A. M. Dirac and the Beauty of Physics." *Scientific American*, May 1, 1993. https://www.scientificamerican.com/article/pam-dirac-and-the-beauty-of-physics.

Cottingham, John. *In Search of the Soul: A Philosophical Essay*. Princeton: Princeton University Press, 2020.

———. *The Spiritual Dimension: Religion, Philosophy, and Human Value*. Cambridge: Cambridge University Press, 2005.

Cottrell, Alan H., and David G. Pettifor. "Models of Structure." In *Structure in Science and Art*, edited by Wendy Pullan and Harshad Bhadeshia, 33–47. Cambridge: Cambridge University Press, 2000.

Coyne, George V., and Michał Heller. *A Comprehensible Universe: The Interplay of Science and Theology*. New York: Springer-Verlag, 2008.

Crick, Francis. *What Mad Pursuit: A Personal View of Scientific Discovery*. New York: Basic, 1988.

Crowther, Paul. *The Kantian Aesthetic: From Knowledge to the Avant-Garde*. Oxford: Oxford University Press, 2014.

Currie, Gregory, et al., eds. *Aesthetics and the Sciences of the Mind*. Oxford: Oxford University Press, 2014.

Davidson, Donald. "Actions, Reasons, and Causes." *Journal of Philosophy* 60 (1963) 685–700.

Davies, E. Brian. *Why Beliefs Matter: Reflections on the Nature of Science*. Oxford: Oxford University Press, 2010.

Davison, Andrew. *Participation in God: A Study in Christian Doctrine and Metaphysics*. Cambridge: Cambridge University Press, 2019.

Dawes, Gregory W. "In Defence of Naturalism." *International Journal for Philosophy of Religion* 70 (2011) 3–25.

———. *Theism and Explanation*. London: Routledge, 2015.

Deacon, Terrence W. *Incomplete Nature: How Mind Emerges from Matter*. New York: Norton, 2013.

de Caro, Mario, and David Macarthur, eds. *Naturalism in Question*. Cambridge: Harvard University Press, 2004.

de Cruz, Helen, and Johan de Smedt. *A Natural History of Natural Theology: The Cognitive Science of Theology and Philosophy of Religion*. Cambridge: MIT Press, 2015.

Dennett, Daniel C. "Aching Voids and Making Voids: A Review of *Incomplete Nature: How Mind Emerged from Matter* by Terrence Deacon." *Quarterly Review of Biology* 88 (2013) 321–24.

———. *The Intentional Stance*. Cambridge: MIT Press, 1989.

de Regt, Henk W. "Beauty in Physical Science Circa 2000." *International Studies in the Philosophy of Science* 16 (2002) 95–103.

De Ridder, Jeroen, et al., eds. *Scientism: Prospects and Problems*. Oxford: Oxford University Press, 2018.

Descartes, René. *Meditations and Other Metaphysical Writings*. Translated by Desmond M. Clarke. London: Penguin Classics, 1998.

de Vries, Paul. "Naturalism in the Natural Sciences." *Christian Scholars Review* 15 (1986) 388–96.

Dicker, Georges. *Descartes: An Analytic and Historical Introduction*. Oxford: Oxford University Press, 2013.

Dirac, P. A. M. "The Evolution of the Physicists' Picture of Nature." *Scientific American*, May 1963, 45–53.

———. "Why We Believe in the Einstein Theory." In *Symmetries in Science*, edited by Bruno Gruber and Richard S. Millman, 1–11. New York: Plenum, 1980.

D'Oro, Giuseppina, and Constantine Sandis, eds. *Reasons and Causes: Causalism and Anti-Causalism in the Philosophy of Action*. London: Palgrave Macmillan, 2013.

Draper, Paul. "Seeking but Not Believing." In *Divine Hiddenness: Essays*, edited by Daniel Howard-Snyder and Paul K. Moser, 197–214. Cambridge: Cambridge University Press, 2001.

Dunshirn, Alfred. "Physis." In *Online Encyclopedia Philosophy of Nature*, edited by Thomas Kirchhoff. Heidelberg: Heidelberg University Press, 2019. https://doi.org/10.11588/oepn.2019.0.66404.

Dupré, John. *The Disorder of Things: Metaphysical Foundations of the Disunity of Science*. Cambridge: Harvard University Press, 1995.

du Sautoy, Marcus. "Exploring the Mathematical Library of Babel." In *Meaning in Mathematics*, edited by John Polkinghorne, 17–26. Oxford: Oxford University Press, 2011.

———. "Grand Designs." In "Infinity and Beyond," edited by Graham Lawton. Special issue, *New Scientist: Infinity and Beyond* 4.4 (2017) 92–95.

Ecklund, Elaine Howard. *Science vs. Religion: What Scientists Really Think*. Oxford: Oxford University Press, 2010.

Eco, Umberto. *The Aesthetics of Thomas Aquinas*. Cambridge: Harvard University Press, 1998.

Eddington, Arthur S. *Mathematical Theory of Relativity*. Cambridge: Cambridge University Press, 1923.

———. *The Nature of the Physical World*. New York: Macmillan, 1928. Originally delivered as the 1927 Gifford Lectures.

Ellis, Fiona. *God, Value, and Nature*. Oxford: Oxford University Press, 2014.

———, ed. *New Models of Religious Understanding*. Oxford: Oxford University Press, 2018.

Elpidorou, Andreas. "The Character of Physicalism." *Topoi* 37 (2018) 435–55. https://link.springer.com/article/10.1007/s11245-017-9488-2.

Engler, Gideon. "Aesthetics in Science and Art." *British Journal of Aesthetics* 30 (1990) 24–34.

———. "Einstein and the Most Beautiful Theories in Physics." *International Studies in the Philosophy of Science* 16 (2002) 27–37.

Evans, C. Stephen. *Preserving the Person: A Look at the Human Sciences*. Vancouver: Regent College, 2002.

———. *Natural Signs and Knowledge of God: A New Look at Theistic Arguments*. Oxford: Oxford University Press, 2010.

Farmelo, Graham. *The Strangest Man: The Hidden Life of Paul Dirac, Quantum Genius*. London: Faber & Faber, 2010.

———. *The Universe Speaks in Numbers: How Modern Maths Reveals Nature's Deepest Secrets*. London: Faber & Faber, 2019.

Feser, Edward. *Aristotle's Revenge: The Metaphysical Foundations of Physical and Biological Science*. Neunkirchen-Seelscheid: Editiones Scholasticae, 2019.

———. *Five Proofs of the Existence of God*. San Francisco: Ignatius, 2017.

———. *The Last Superstition: A Refutation of the New Atheism*. South Bend, IN: St. Augustine's, 2008.

Ford, James L. *The Divine Quest, East and West: A Comparative Study of Ultimate Realities*. Albany: State University of New York Press, 2016.

Forte, Bruno. *The Portal of Beauty: Towards a Theology of Aesthetics*. Grand Rapids: Eerdmans, 2008.

Frame, John. *Nature's Case for God: A Brief Biblical Argument*. Bellingham, WA: Lexham, 2018.

Franklin, James. *An Aristotelian Realist Philosophy of Mathematics: Mathematics as the Science of Quantity and Structure*. Basingstoke, UK: Palgrave Macmillan, 2014.

———. "Uninstantiated Properties and Semi-Platonist Aristotelianism." *Review of Metaphysics* 69 (2015) 25–45.

French, Steven. *The Structure of the World: Metaphysics and Representation*. Oxford: Oxford University Press, 2014.

Friedman, Michael. "Exorcising the Philosophical Tradition: Comments on John McDowell's *Mind and World*." *Philosophical Review* 105 (1996) 427–67.

Gabbani, Chiara, ed. "Between Two Images: The Manifest and the Scientific Conceptions of the Human Being." Special issue, *HumanaMente: Quarterly Journal of Philosophy* 21 (2012).

Gal, Ofer. *The Origins of Modern Science: From Antiquity to the Scientific Revolution*. Cambridge: Cambridge University Press, 2021.

Galilei, Galileo. *The Assayer*. In *Discoveries and Opinions of Galileo*, translated by Stillman Drake, 231–80. New York: Anchor, 1957.

Gasser, Georg, ed. *How Successful Is Naturalism?* Heusenstamm: Ontos-Verlag, 2007.

Gell-Mann, Murray. "Banquet Speech." Nobel Banquet, Stockholm, Dec. 10, 1969. https://www.nobelprize.org/prizes/physics/1969/gell-mann/speech.

Genova, A. C. Review of *Reading McDowell on Mind and World*, edited by Nicholas S. Smith. *Notre Dame Philosophical Reviews* (2003). https://ndpr.nd.edu/reviews/reading-mcdowell-on-mind-and-world.

Gerber, Daniel. "Descartes, Mechanics and Mechanical Philosophy." *Midwest Studies in Philosophy* 26 (2002) 185–204.

Gerth, H. H., and C. Wright Mills, trans. and eds. *From Max Weber: Essays in Sociology*. New York: Oxford University Press, 1946.

Gill, Eric. *Beauty Looks After Itself*. With an introduction by Catherine Pickstock. Rancho de Taos, NM: Sophia Perennis, 2012.

Gimbel, Steven. *Einstein: His Space and Times*. New Haven: Yale University Press, 2015.

Gingerich, Owen. "Dare a Scientist Believe in Design?" In *God's Universe*, 43–81. Cambridge: Belknap, 2006.

Glock, Hans-Johann, ed. *Wittgenstein: A Critical Reader*. Oxford: Blackwell, 2001.

Glynn, Ian. *Elegance in Science: The Beauty of Simplicity*. Oxford: Oxford University Press, 2010.

Goetz, Stewart, and Charles Taliaferro. *Naturalism*. Grand Rapids: Eerdmans, 2008.

Gould, Carol S. "The Reality of Aesthetic Properties: A Response to Goldman." *Journal of Aesthetics and Art Criticism* 52 (1994) 349–51.

Grant, Edward. *A History of Natural Philosophy: From the Ancient World to the Nineteenth Century*. Cambridge: Cambridge University Press, 2007.

Gray, Asa. *The Letters of Asa Gray*. 2 vols. Boston and New York: Houghton, Mifflin, 1896. Reprint, Cambridge: Cambridge Library Collection, 2015.

Gregersen, Niels Henrik, et al., eds. *The Concept of Nature in Science and Theology, Part II*. Vol. 4 of *Studies in Science and Theology*. Geneva: Labor et Fides, 1996.

Gribbin, John, and Mary Gribbin. *Out of the Shadow of a Giant: Hooke, Halley and the Birth of British Science*. London: William Collins, 2017.

Groothuis, Douglas. "Metaphysical Implications of Cosmological Arguments: Exorcising the Ghost of Hume." In Sennett and Groothuis, *In Defense of Natural Theology*, 107–22.

Gross, David J. "Physics and Mathematics at the Frontier." *Proceedings of the National Academy of Sciences of the United States of America* 85 (1988) 8371–75.

Guaspari, David. "The Sum of Its Parts." Review of *An Aristotelian Realist Philosophy of Mathematics*, by James Franklin. *New Criterion* 33 (2014) 62–65.

Hacker, Peter M. S. "Strawson's Concept of a Person." *Proceedings of the Aristotelian Society* 102 (2002) 21–40.

Hacker, Peter M. S., and M. R. Bennett. *Philosophical Foundations of Neuroscience.* Oxford: Blackwell, 2003.

Halvorson, Hans. "Why Methodological Naturalism?" In *The Blackwell Companion to Naturalism*, edited by Kelly James Clark, 136–49. Oxford: Wiley-Blackwell, 2014.

Hanna, Robert. *Kant, Science, and Human Nature.* Oxford: Oxford University Press, 2006.

Hannam, James. *The Genesis of Science: How the Christian Middle Ages Launched the Scientific Revolution.* Washington, DC: Regnery, 2009.

———. *God's Philosophers: How the Medieval World Laid the Foundations of Modern Science.* London: Icon, 2009.

Harman, Gilbert, and Ernest Lepore, eds. *A Companion to W. V. O. Quine.* Wiley-Blackwell Companions to Philosophy. Oxford: Wiley-Blackwell, 2014.

Harries, Richard. *Art and the Beauty of God: A Christian Understanding.* London: Mowbray, 1993.

Harrison, Peter. *The Territories of Science and Religion.* Chicago: University of Chicago Press, 2015.

Harrison, Peter, and Jon H. Roberts, eds. *Science Without God? Rethinking the History of Scientific Naturalism.* Oxford: Oxford University Press, 2019.

Hart, David Bentley. "Beauty, Being, Kenosis." In *Theological Territories*, 245–56. Notre Dame: University of Notre Dame Press, 2020.

———. *The Beauty of the Infinite: The Aesthetics of Christian Truth.* Grand Rapids: Eerdmans, 2003.

———. *That All Shall Be Saved: Heaven, Hell, and Universal Salvation.* New Haven: Yale University Press, 2019.

Hattab, Helen. *Descartes on Forms and Mechanisms.* Cambridge: Cambridge University Press, 2011.

Heisenberg, Werner. *Across the Frontiers.* New York: Harper & Row, 1974.

Heller, Michael. "Science and Transcendence." *Studies in Science and Theology* 4 (1996) 3–12.

Hick, John. *Philosophy of Religion.* Englewood Cliffs, NJ: Prentice Hall, 1975.

Hincks, Adam D. "What Does Physical Cosmology Say About Creation from Nothing?" In *Creation ex Nihilo: Origins, Development, Contemporary Challenges*, edited by Gary A. Anderson and Markus Bockmuehl, 319–46. Notre Dame: University of Notre Dame Press, 2018.

Hobart, Mark. *The Great Rift: Literacy, Numeracy, and the Religion–Science Divide.* Cambridge: Harvard University Press, 2018.

Holder, Rodney. *Natural Theology and the Legacy of Karl Barth.* West Conshohocken, PA: Templeton, 2012.

Holton, Gerald. *Thematic Origins of Scientific Thought: Kepler to Einstein.* Cambridge: Harvard University Press, 1973.

Hookway, Christopher. *Quine.* Key Thinkers Series. Oxford: Polity, 1988.

Hooper, Dan. *At the Edge of Time: Exploring the Mysteries of Our Universe's First Few Seconds.* Princeton: Princeton University Press, 2019.

Hossenfelder, Sabine. "Beauty Is Truth, Truth Is Beauty, and Other Lies of Physics." Edited by Sally Davies. *Aeon*, July 11, 2018. https://aeon.co/ideas/beauty-is-truth-truth-is-beauty-and-other-lies-of-physics.

———. *Lost in Math: How Beauty Leads Physics Astray.* New York: Basic, 2018.

Hume, David. *A Treatise of Human Nature*. 1739–1740. Edited by L. A. Selby-Bigge. Oxford: Oxford University Press, 1978.

Hunter, James Davison, and Paul Nedelisky. *Science and the Good: The Tragic Quest for the Foundations of Morality*. New Haven: Yale University Press, 2018.

Hutton, Ronald. *Pagan Britain*. New Haven: Yale University Press, 2013.

Irenaeus. *Against Heresies*. Translated by Dominic J. Unger and John J. Dillon. Mahwah, NJ: Newman, 2012.

Ivanova, Milena. "Poincaré's Aesthetics of Science." *Synthese* 194 (2017) 2581–94. https://www.jstor.org/stable/48726139.

Ivanova, Milena, and Steven French, eds. *The Aesthetics of Science: Beauty, Imagination and Understanding*. New York: Routledge, 2020.

Jaeger, Lydia. *What the Heavens Declare: Science in the Light of Creation*. Eugene, OR: Cascade, 2012.

Johnson, Dominic, and Jesse Bering. "Hand of God, Mind of Man." In *The Believing Primate: Scientific, Philosophical, and Theological Reflections on the Origin of Religion*, edited by Jeffrey Schloss and Michael J. Murray, 26–44. Oxford: Oxford University Press, 2009.

Kantzer Komline, Han-Luen. Review of *The Beauty of the Infinite: The Aesthetics of Christian Truth*, by David Bentley Hart. *Princeton Theological Review* 12 (2006) 39–40.

Keas, Michael N. *Unbelievable: 7 Myths About the History and Future of Science and Religion*. Wilmington, DE: ISI, 2019.

Kim, Jaegwon. "The American Origins of Philosophical Naturalism." In "Philosophy in America at the Turn of the Century," edited by Robert Audi. Special issue, *Journal of Philosophical Research* 28 (2003) 83–98.

Kincaid, Harold, et al., eds. *Value-Free Science? Ideals and Illusions*. Oxford: Oxford University Press, 2007.

Kohn, Livia. "Ultimate Reality: Chinese Religion." In *Ultimate Realities*, edited by Robert C. Neville, 9–35. Albany: State University of New York Press, 2001.

Koyré, Alexandre. *From the Closed World to the Infinite Universe*. Eugene, OR: Angelico, 2016.

Kragh, Helge. *Dirac: A Scientific Biography*. Cambridge: Cambridge University Press, 1990.

———. "Mathematics and Physics: The Idea of a Pre-Established Harmony." *Science & Education* 24 (2015) 515–27.

———. "Paul Dirac: Seeking Beauty." *Physics World* 15.8 (2002) 27–31.

Krauss, Lawrence M. *A Universe from Nothing: Why There Is Something Rather than Nothing*. London: Simon & Schuster, 2012.

Ladyman, James. "Structural Realism." From *The Stanford Encyclopedia of Philosophy* (Winter 2020 ed.), edited by Edward N. Zalta. https://plato.stanford.edu/archives/win2020/entries/structural-realism.

———. "What Is Structural Realism?" *Studies in History and Philosophy of Science* 29 (1998) 409–24.

Larvor, Brendan. "Naturalism." In *The Wiley Handbook of Humanism*, edited by Andrew Copson and A. C. Grayling, 37–54. London: Wiley-Blackwell, 2015.

Lawrence, Sarah, and Mark McCartney, eds. *Mathematicians and Their Gods: Interactions Between Mathematics and Religious Beliefs*. Oxford: Oxford University Press, 2015.

Lederman, Leon M., and Christopher T. Hill. *Symmetry and the Beautiful Universe.* Amherst, NY: Prometheus, 2008.

Legenhausen, Gary. "Is God a Person?" *Religious Studies* 22 (1986) 307–23.

Lennox, James G. "The Darwin/Gray Correspondence 1857–1869: An Intelligent Discussion About Chance and Design." *Perspectives on Science* 18 (2010) 456–79.

———. "Robert Boyle's Defense of Teleological Inference in Experimental Science." *Isis* 74 (1983) 38–52.

Leunissen, Mariska, ed. *Aristotle's Physics: A Critical Guide.* Cambridge Critical Guides. Cambridge: Cambridge University Press, 2018.

Levinson, Jerrold, ed. *The Oxford Handbook of Aesthetics.* Oxford: Oxford University Press, 2005.

Lewis, Geraint F., and Luke A. Barnes. *A Fortunate Universe: Life in a Finely Tuned Cosmos.* Cambridge: Cambridge University Press, 2016.

Lindgaard, Jakob, ed. *John McDowell: Experience, Norm, and Nature.* Oxford: Blackwell, 2008.

Livio, Mario. *The Accelerating Universe: Infinite Expansion, the Cosmological Constant, and the Beauty of the Cosmos.* New York: Wiley, 2000.

Lloyd, G. E. R. *Being, Humanity, and Understanding.* Oxford: Oxford University Press, 2012.

Locke, John. *An Essay Concerning Human Understanding.* Edited by P. H. Nidditch. Clarendon Edition of the Works of John Locke. Oxford: Oxford University Press, 1979.

Long, A. A. *Greek Models of Mind and Self.* Cambridge: Harvard University Press, 2015.

Lucretius. *On the Nature of the Universe.* Translated by Ronald Melville. Oxford World's Classics. Oxford: Oxford University Press, 1999.

Magee, Bryan. *Philosophy and the Real World: An Introduction to Karl Popper.* Chicago: Open Court, 1985.

Maitzen, Stephen. "Against Ultimacy." In *Renewing Philosophy of Religion: Exploratory Essays*, edited by Paul Draper and J. L. Schellenberg, 48–63. Oxford: Oxford University Press, 2017.

Manson, Neil, ed. *God and Design: The Teleological Argument and Modern Science.* London: Routledge, 2003.

Maxwell, Nicholas. *The Comprehensibility of the Universe: A New Conception of Science.* Oxford: Oxford University Press, 2003.

McAllister, James W. "Dirac and the Aesthetic Evaluation of Theories." *Journal of Methodology and Science* 23 (1990) 87–102.

———. "Recent Work on Aesthetics and Science." *International Studies in the Philosophy of Science* 16 (2002) 7–11.

McCubbin, N. A. "Beauty in Physics: The Legacy of Paul Dirac." *Contemporary Physics* 45 (2004) 319–33.

McDonnell, Jane. *The Pythagorean World: Why Mathematics Is Unreasonably Effective in Physics.* London: Palgrave Macmillan, 2017.

McDowell, John. *Mind and World.* Cambridge: Harvard University Press, 1996.

McGrath, Alister E. *A Fine-Tuned Universe: The Quest for God in Science and Theology.* Louisville: Westminster John Knox, 2009.

McLeish, Tom. *The Poetry and Music of Science: Comparing Creativity in Science and Art.* Oxford: Oxford University Press, 2019.

Meisner, Gary B. *The Golden Ratio: The Divine Beauty of Mathematics*. Minneapolis: Race Point, 2018.

Meister, Chad, and James K. Dew Jr. *God and Evil: The Case for God in a World Filled with Pain*. Downers Grove, IL: IVP Academic, 2013.

Milbank, John. "Beauty and the Soul." In *Theological Perspectives on God and Beauty*, edited by John Milbank et al., 1–34. London and New York: Trinity Press International, 2003.

Millican, Peter. "Hume's Fork and His Theory of Relations." *Philosophy and Phenomenological Research* 95 (2017) 3–65.

Molland, A. G. "The Geometrical Background to the 'Merton School.'" *British Journal for the History of Science* 4 (1968) 108–25.

Montano, Uskali. "Beauty in Science: A New Model of the Role of Aesthetic Evaluations in Science." *European Journal for Philosophy of Science* 3 (2013) 133–56.

Moreland, J. P., et al., eds. *Debating Christian Theism*. Oxford: Oxford University Press, 2013.

Morris, William Edward, and Charlotte Brown. "David Hume." From *The Stanford Encyclopedia of Philosophy* (Spring 2021 ed.), edited by Edward N. Zalta. https://plato.stanford.edu/archives/spr2021/entries/hume.

Nagel, Thomas. *The Last Word*. Oxford: Oxford University Press, 1997.

———. *The View from Nowhere*. Oxford: Oxford University Press, 1989.

New Scientist. "News in Brief." July 11, 2020, 18–19.

Nichols, Aidan. *A Key to Balthasar: Hans Urs von Balthasar on Beauty, Goodness, and Truth*. Grand Rapids: Baker Academic, 2011.

North, Jill. "The 'Structure' of Physics: A Case Study." *Journal of Philosophy* 106 (2009) 57–88.

O'Connor, Timothy. *Theism and Ultimate Explanation: The Necessary Shape of Contingency*. Chichester, UK: Wiley-Blackwell, 2012.

O'Hear, Anthony. "Science and Religion." *British Journal for the Philosophy of Science* 44 (1993) 505–16.

Onyango Okello, Joseph B. *A History and Critique of Methodological Naturalism: The Philosophical Case for God's Design of Nature*. Eugene, OR: Wipf & Stock, 2016.

Paley, William. *Natural Theology: Or, Evidence of the Existence and Attributes of the Deity, Collected from the Appearances of Nature*. Oxford World's Classics. Oxford: Oxford University Press, 2008.

Panchuk, Michelle. Review of *The Hiddenness of God*, by Michael C. Rea. *Faith and Philosophy* 36 (2019) 280–85.

Papineau, David. *Philosophical Naturalism*. Oxford: Wiley-Blackwell, 1993.

———. *The Roots of Reason: Philosophical Essays on Rationality, Evolution and Probability*. Oxford: Oxford University Press, 2006.

Parsons, Glenn, and Allen Carlson. "Critical Notice: Eddy Zemach, *Real Beauty*." *Canadian Journal of Philosophy* 29 (1999) 635–54.

———. *Functional Beauty*. Oxford: Oxford University Press, 2008.

Parsons, Keith. *It Started with Copernicus: Vital Questions About Science*. Amherst, NY: Prometheus, 2014.

Plantinga, Alvin. *Warranted Christian Belief*. Oxford: Oxford University Press, 2000.

———. *Where the Conflict Really Lies: Science, Religion, and Naturalism*. Oxford: Oxford University Press, 2012.

Plato. *Timaeus and Critias*. Translated by Robin Waterfield. Oxford World's Classics. Oxford: Oxford University Press, 2008.

Polkinghorne, John. "Mathematical Reality." In *Meaning in Mathematics*, 27–35. Cambridge: Cambridge University Press, 2011.

Popper, Karl. *The Logic of Scientific Discovery*. 2nd ed. London: Routledge Classics, 2002.

Psillos, Stathis. *Knowing the Structure of Nature: Essays on Realism and Explanation*. Basingstoke, UK: Palgrave Macmillan, 2009.

Pullan, Wendy. "Introduction: Structuring Structure." In *Structure in Science and Art*, edited by Wendy Pullan and Harshad Bhadeshia, 1–8. Cambridge: Cambridge University Press, 2000.

Putnam, Hilary. *The Collapse of the Fact/Value Dichotomy and Other Essays*. Cambridge: Harvard University Press, 2002.

———. *Naturalism, Realism, and Normativity*. Cambridge: Harvard University Press, 2016.

Quinton, Anthony. *From Wodehouse to Wittgenstein: Essays*. Manchester: Carcanet, 1998.

Ramirez, Jorge R. E. "Augustine's Proof for God's Existence from the Experience of Beauty." *Augustinian Studies* 19 (1988) 121–30.

Rasmussen, Joshua, and Felipe Leon. *Is God the Best Explanation of Things? A Dialogue*. Cham, Switzerland: Palgrave Macmillan, 2019.

Rea, Michael C. "Divine Presence in a Material World." In *Hiddenness of God*, 113–36.

———. *The Hiddenness of God*. Oxford: Oxford University Press, 2018.

Reichenbach, Hans. *Experience and Prediction*. Chicago: University of Chicago Press, 1938.

Reid, Thomas. *Lectures on Natural Theology*. Edited by James A. Barham and Jonathan Akins. Nashville: Influence, 2020.

Reppert, Victor. *C. S. Lewis's Dangerous Idea: In Defense of the Argument from Reason*. Downers Grove, IL: InterVarsity, 2003.

Resnik, Michael D. *Mathematics as a Science of Patterns*. Oxford: Oxford University Press, 1997.

Ridley, Aaron. "Expression in Art." In *The Oxford Handbook of Aesthetics*, edited by Jerrold Levinson, 211–27. Oxford: Oxford University Press, 2003.

Rist, John M. *What Is a Person? Realities, Constructs, Illusions*. Cambridge: Cambridge University Press, 2020.

Roberts, Louis. *The Theological Aesthetics of Hans Urs von Balthasar*. Washington, DC: Catholic University of America Press, 1987.

Robinson, Daniel, ed. *Neuroscience and Philosophy: Mind, Brain, and Language*. New York: Columbia University Press, 2007.

Rosch, Eleanor H. "Cognitive Reference Points." *Cognitive Psychology* 7 (1975) 532–47.

———. "Cognitive Representations of Semantic Categories." *Journal of Experimental Psychology* 104 (1975) 192–233.

———. "Natural Categories." *Cognitive Psychology* 4 (1973) 328–50.

Rovelli, Carlo. *Helgoland: The Strange and Beautiful Story of Quantum Physics*. London: Penguin Random House, 2020.

———. *Seven Brief Lessons on Physics*. Anniversary ed. London: Allen Lane, 2025.

Rundle, Bede. "Meaning and Understanding." In Glock, *Wittgenstein*, 94–99.

Ruse, Michael. "The Argument from Design." In *Debating Design: From Darwin to DNA*, edited by William Dembski and Michael Ruse, 13–31. Cambridge: Cambridge University Press, 2004.

———. *Evolution and Religion: A Dialogue*. London: Rowman & Littlefield, 2016.

———. *Science and Spirituality: Making Room for Faith in the Age of Science*. Cambridge: Cambridge University Press, 2020.

Schindler, D. C. *Love and the Postmodern Predicament: Rediscovering the Real in Beauty, Goodness, and Truth*. Eugene, OR: Cascade, 2018.

Schneider, Susan. "Does the Mathematical Nature of Physics Undermine Physicalism?" *Journal of Consciousness Studies* 24.9–10 (2017) 7–39.

Schrödinger, Erwin. *Nature and the Greeks and Science and Humanism*. Cambridge: Cambridge University Press, 2014.

Schroeder, Severin. *Wittgenstein*. Cambridge: Polity, 2006.

Sehon, Scott. *Teleological Realism: Mind, Agency, and Explanation*. Cambridge: MIT Press, 2005.

Sellars, Wilfrid. *Science, Perception and Reality*. New York: Ridgeview, 2017.

Sennett, James F. "Hume's Stopper and the Natural Theology Project." In Sennett and Groothuis, *In Defense of Natural Theology*, 82–106.

Sennett, James F., and Douglas Grootuis, eds. *In Defense of Natural Theology: A Post-Humean Assessment*. Downers Grove, IL: InterVarsity, 2005.

Sevier, Christopher Scott. *Aquinas on Beauty*. Lanham, MD: Lexington, 2015.

Shanks, Niall. *God, the Devil, and Darwin: A Critique of Intelligent Design Theory*. Oxford: Oxford University Press, 2004.

Shapiro, Stewart. *Thinking About Mathematics: The Philosophy of Mathematics*. Oxford: Oxford University Press, 2011.

Sherry, Patrick. *Spirit and Beauty: An Introduction to Theological Aesthetics*. London: SCM, 2002.

Slagle, James. *The Epistemological Skyhook: Determinism, Naturalism and Self-Defeat*. New York: Routledge, 2016.

Smith, Nicholas H., ed. *Reading McDowell on Mind and World*. London: Routledge, 2002.

Snell, Bruno. *The Discovery of the Mind in Greek Philosophy and Literature*. New York: Dover, 1982.

Snell, Richard J., and Steven F. McGuire, eds. *Concepts of Nature: Ancient and Modern*. Lanham, MD: Lexington, 2016.

Sober, Elliott. *The Design Argument*. Cambridge: Cambridge University Press, 2018.

Southgate, Christopher. *God, Humanity, and the Cosmos*. 3rd ed. London: T&T Clark, 2011.

Stanley, Matthew. *Practical Mystic: Religion, Science and A. S. Eddington*. Chicago: University of Chicago Press, 2007.

Steane, Andrew. *Science and Humanity: A Humane Philosophy of Science and Religion*. Oxford: Oxford University Press, 2018.

Stenger, Victor J. "The Universe Shows No Evidence of Design." In *Debating Christian Theism*, edited by J. P. Moreland et al., 47–60. Oxford: Oxford University Press, 2013.

Stenmark, Mikael. *How to Relate Science and Religion: A Multidimensional Model*. Grand Rapids: Eerdmans, 2005.

Stevenson, Leslie. *Open to New Light: Quaker Spirituality in Historical and Philosophical Context*. Exeter, UK: Imprint Academic, 2012.

Stipp, David. *A Most Elegant Equation: Euler's Formula and the Beauty of Mathematics*. New York: Basic, 2017.

Stone, Jerome A. *Religious Naturalism Today*. Albany: State University of New York Press, 2009.

Strawson, P. F. *Individuals: An Essay in Descriptive Metaphysics*. London: Methuen, 1959.

Suppes, Patrick. "A Comparison of the Meaning and Uses of Models in Mathematics and the Empirical Sciences." *Synthese* 12 (1960) 287–301.

Swinburne, Richard. *The Coherence of Theism*. Oxford: Clarendon, 1977.

———. "Discussion: Reply to Grünbaum." *British Journal for the Philosophy of Science* 51 (2000) 481–85.

Taliaferro, Charles. *Aesthetics: A Beginner's Guide*. Oxford: Oneworld, 2011.

Taliaferro, Charles, and Chad Meister. *Contemporary Philosophical Theology*. New York: Routledge, 2016.

Tallon, Philip. "The Mozart Argument and the Argument for Play and Enjoyment: The Theistic Argument from Beauty and Play." In *The Plantinga Project: Two Dozen (or So) Arguments for God*, edited by Jerry L. Walls and Trent Dougherty, 321–40. Oxford: Oxford University Press, 2018.

Tegmark, Max. *Our Mathematical Universe: My Quest for the Ultimate Nature of Reality*. London: Penguin, 2015.

Thornton, Tim. "Experience, Knowledge, and Openness to the World." In *John McDowell*, 177–207. 2nd ed. Oxford: Routledge, 2019.

Travis, Melissa Cain. "A Glorious Resonance: The Intelligibility of Nature and the *Imago Dei*." In *The Story of the Cosmos: How the Heavens Declare the Glory of God*, edited by Paul M. Gould and Daniel Ray, 21–36. Eugene, OR: Harvest House, 2019.

van Inwagen, Peter. *The Problem of Evil*. Oxford: Oxford University Press, 2006.

———. "Quam Dilecta." In *God and the Philosophers: The Reconciliation of Faith and Reason*, edited by Thomas V. Morris, 31–60. Oxford: Oxford University Press, 1995.

———. "Russell's China Teapot." In *The Right to Believe: Perspectives in Religious Epistemology*, edited by Dariusz Łukasiewicz and Roger Pouivet, 11–26. Frankfurt: Ontos-Verlag, 2012.

van Inwagen, Peter, and Dean Zimmerman. *Persons: Human and Divine*. Oxford: Oxford University Press, 2007.

Vernon, Mark. *A Secret History of Christianity: Jesus, the Last Inkling, and the Evolution of Consciousness*. Alresford, UK: John Hunt, 2019.

Viladesau, Richard. *Theological Aesthetics: God in Imagination, Beauty, and Art*. Oxford: Oxford University Press, 1999.

von Balthasar, Hans Urs. *The Glory of the Lord: A Theological Aesthetic*. San Francisco: Ignatius, 1982–1991.

———. *Theo-Drama: Theological Dramatic Theory*. San Francisco: Ignatius, 1988–1998.

———. *Theo-Logic: Theological Logical Theory*. San Francisco: Ignatius, 2000–2005.

———. "Transcendentality and Gestalt." *Communio* 11 (1984) 29–39.

Wagner, Richard, and Andrew Briggs. *The Penultimate Curiosity: How Science Swims in the Stream of Ultimate Questions*. Oxford: Oxford University Press, 2016.

Wahlberg, Mats. *Reshaping Natural Theology: Seeing Nature as Creation.* London: Palgrave Macmillan, 2012.

Wainwright, Geoffrey. Review of *The Beauty of the Infinite: The Aesthetics of Christian Truth. First Things*, Dec. 23, 2003. https://www.firstthings.com/article/2004/03/the-beauty-of-the-infinite.

Ward, Keith. *Sharing in the Divine Nature: A Personalist Metaphysics.* Eugene, OR: Cascade, 2020.

Waterlow, Sarah. *Nature, Change, and Agency in Aristotle's Physics: A Philosophical Study.* Oxford: Clarendon, 1988.

Weber, Max. "Science as a Vocation." In Gerth and Mills, *From Max Weber*, 129–56.

Wechsler, Judith. *On Aesthetics in Science.* Cambridge: MIT Press, 1987.

Weinberg, Steven. *Dreams of a Final Theory.* New York: Vintage, 1993.

White, Roger M. *Talking About God: The Concept of Analogy and the Problem of Religious Language.* Farnham, UK: Ashgate, 2010.

Wiggins, David. "Cognitivism, Naturalism and Normativity: A Reply to Peter Railton." In *Reality, Representation, and Projection*, edited by John Haldane and Crispin Wright, 293–327. Oxford: Oxford University Press, 1993.

———. "A Neglected Position?" In *Reality, Representation, and Projection*, edited by John Haldane and Crispin Wright, 328–38. Oxford: Oxford University Press, 1993.

Wigner, Eugene. "The Unreasonable Effectiveness of Mathematics in the Natural Sciences." *Communications in Pure and Applied Mathematics* 13 (1960) 1–14.

Wilczek, Frank. *A Beautiful Question: Finding Nature's Deep Design.* London: Penguin, 2015.

———. "When Words Fail." *Nature* 410 (2001) 149. https://doi.org/10.1038/35065756.

Wildman, Wesley J. "Comparative Natural Theology." *American Journal of Theology & Philosophy* 27 (2006) 173–99. https://www.jstor.org/stable/27944377.

———. "Really Ultimate Reality." *Theology and Science* 15 (2017) 260–65.

Willard, Dallas. "The Three-Stage Argument for the Existence of God." In *Contemporary Perspectives on Religious Epistemology*, edited by R. Douglas Geivett and Brendan Sweetman, 21–42. New York: Oxford University Press, 1992.

Williams, Bernard. *Descartes: The Project of Pure Enquiry.* London: Routledge Classics, 2014.

Williams, Peter S. *C. S. Lewis and the New Atheists.* Milton Keynes: Paternoster, 2013.

Williams, Robert N., and Daniel N. Robinson. *Scientism: The New Orthodoxy.* London: Bloomsbury, 2015.

Wolfram, Stephen. *A New Kind of Science.* Champaign, IL: Wolfram Media, 2002.

Worrall, John. "Structural Realism: The Best of Both Worlds?" *Dialectica* 43 (1989) 99–124.

Zangwill, Nick. "Aesthetic Realism." In *The Oxford Handbook of Aesthetics*, edited by Jerrold Levinson, 63–78. Oxford: Oxford University Press, 2005.

Zee, A. *Fearful Symmetry: The Search for Beauty in Modern Physics.* Princeton: Princeton University Press, 2007.

Zemach, Eddy M. "Properties, Aesthetic Laws, and Aesthetic Principles." *Journal of Aesthetics and Art Criticism* 46 (1987) 67–73.

———. "Real Beauty." *Midwest Studies in Philosophy* 16 (1991) 249–65.

———. *Real Beauty.* University Park: Pennsylvania State University Press, 1997.

Index

www.ingramcontent.com/pod-product-compliance
Lightning Source LLC
LaVergne TN
LVHW050806061126
851857LV00012B/703